T0222408

The UFO Phenomenon and The Origin Of Mass Extinctions

ROBERT ITURRALDE

BALBOA.PRESS
A DIVISION OF HAY HOUSE

Balboa Press books may be ordered through booksellers or by contacting:

Balboa Press
A Division of Hay House
1663 Liberty Drive
Bloomington, IN 47403
www.balboapress.com
844-682-1282

Because of the dynamic nature of the Internet, any web addresses or links contained in this book may have changed since publication and may no longer be valid. The views expressed in this work are solely those of the author and do not necessarily reflect the views of the publisher, and the publisher hereby disclaims any responsibility for them.

The author of this book does not dispense medical advice or prescribe the use of any technique as a form of treatment for physical, emotional, or medical problems without the advice of a physician, either directly or indirectly. The intent of the author is only to offer information of a general nature to help you in your quest for emotional and spiritual well-being. In the event you use any of the information in this book for yourself, which is your constitutional right, the author and the publisher assume no responsibility for your actions.

Any people depicted in stock imagery provided by Getty Images are models, and such images are being used for illustrative purposes only. Certain stock imagery © Getty Images.

Print information available on the last page.

ISBN: 979-8-7652-4929-1 (sc)
ISBN: 979-8-7652-4998-7 (e)

Library of Congress Control Number: 2024902107

Balboa Press rev. date: 04/29/2024

CONTENTS

INTRODUCTION

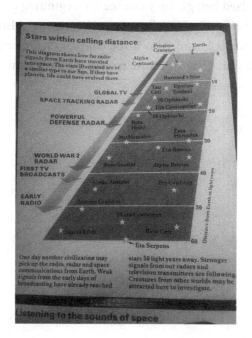

"In any case, the significance of UFOs may be that our fate is in the hands of beings from elsewhere for good or ill. If so, their powers and abilities are being known to us, one might say, but their minds and thoughts remain shrouded in mystery. If UFOs carry advance beings from another civilization in space, they may well be a sword of Damocles poised over our heads".

—Richard Hall, Ufo Scholar 12-25-1930-7-17-2009

Since the early days of radio, radar, and tv broadcasting more than 50 years ago. All these early signs of human civilization are traveling in space. As a result, if they are alien civilizations out there, now they are aware of our existence. Conversely, there are people who claimed that they had been taking on board UFOs. Aliens talk to them about the dangers of nuclear war, and global warming.

Furthermore, aliens told them that they have super-telescopes that they can see houses and apartments and the people in there. Also they can see when planets have life and they manipulate organic evolution and give religion to organized beings. They can see the beginning of life in different planets. I ask if their evolution follow Darwin concepts? Also, if they follow Marx historical materialism? Also if the concept of Einstein relativity are familiar? So, let's wait what the future reserve for the human race.?

Chapter 1

Unexplainable Archaeological Findings before the evolution of Homo Sapiens

"We are sending them out. Why shouldn't someone be sending here? Just think, When pioneer 10 finally leaves our solar system, it becomes our first ufo to all the other star systems"

Over the past several decades, south African miners have found hundreds of metallic spheres Bluish metal with white flecks and another which is a hollow ball filled with a white spongy center. Moreover, Roelf Marx curator of the museum of Klerksdorp in South Africa where some of the spheres are housed said, "the spheres are a complete mystery, they look man-made yet at the time when the

Spheres appear to rest in this rock from the Pre-Cambrian Era, evolution haven't began creating complex organisms like Homo sapiens. The sedimentation where the spheres were found is 2.8 billion years old. The globes which have a fibrous structure on the inside with a shell around it, are very hard and cannot be scratched, even by steel". The Mohs' scale of hardness is named after Friedrich Mohs Who chose 10 minerals as reference points for comparative hardness. The spheres are mysterious because were found in a mineral deposit 2.8 billion years old and it Was not life or complex organisms like Homo sapiens and this is the reason I postulate the theory that

The intelligence behind the ufo phenomenon left all these unexplainable archeological findings for the Human race to see, to wonder, to ask about the possibility of life in the Universe. Furthermore, for the Intelligence behind the ufo phenomenon in the Pre-Cambrian Era this is the best way to leave For posterity the fingerprints of their presence on planet earth.

A shoe sole from Nevada

On October 8, 1922, In Nevada John T Reid a mining engineer stopped suddenly and looked in amazement at a rock near his feet because the rock it seemed to be a human footprint! Actually, looking closely it was not a human feet but a shoe sole which had been turned into stone, the part In the front was missing but there was the outline of at least two-thirds of the shoe sole. Moreover, around the outline it was a well-defined sewn thread which had, it appeared attached the welt to the sole. furthermore, was another line of sewing and in the center, where the foot would have rested had the object been really a shoe sole. In addition, there was an indentation, exactly such as would have been made by the bone of the heel rubbing upon and wearing down the material of which the sole had been made. The shoe sole incase in a Triassic rock is dated 213-248 million years ago in the Triassic era. Again we found ourselves with the unexplainable for any reasonable logic because evolution in the Triassic haven't created Homo sapiens. Further, the primitive man didn't sew not even the near dental man knew how to sew. However, someone left the shoe sole for posterity and the only I can think of is that the ufo phenomenon is responsible. The reason is the only extraterrestrial intelligence that exist in this planet, because they have the technology to travel in time. In fact, there are videos showing ufos going inside volcanoes and going even into the surface of the sun. So, I see the possibility that the Intelligence behind the ufo exist on the planet deep inside earth. We know for sure that the ufo phenomenon have bases in oceans and lakes of the world. So they can have bases inside the earth because their machines and the alien entities can exist in inhospitable places.

As a matter-of-fact, geologists have discover two mysterious structures deep inside the earth, at the boundary between the core and the mantle. The blobs -one under West Africa, the other under the Pacific ocean span an area as large as a continent, and stretch hundreds of miles into the mantle. For example, in 1947 10 people among them was a Dishman in Idaho. A housewife witnessed a fleet of eight ufos in the mountainside near St Maries, Idaho. She said they came into view "At an extreme speed suddenly slow, and then fluttered like leaves to the ground"

The most mysterious part of the story is that when the witnesses were looking at the ground, they could not find them. So the ufos neutralize the electromagnetic forces and then became invisible. The electromagnetic force create the resistance between physical bodies.

Metallic Tubes from Chalk in France,

Y.Druet and H.Salfati in 1968 announced the discovery of semi-ovoid metallic tubes of identical shape but varying in size incrusted in a cretaceous chalk. The chalk bed, exposed in a quarry at Saint Jean de Livet, France it was estimated to be 65 million years old. So they eliminated various theories and they concluded that intelligent beings had lived 65 million years ago. Obviously, we all know that 65 million years ago mammals having evolved. We were waiting in the Russian roulette we call evolution to create the mammals. So I don't have other alternative that postulate my theory that the Intelligence behind the Ufo phenomenon is responsible for these unexplainable archeological findings. Why? The first reason is because I see that ufos can go into volcanoes, deep into the Oceans and lakes, there is evidence that ufos go into the sun's surface. Also ufos can travel in time into the past, future . The intelligence behind the ufo phenomenon they can see the future also they want to leave for posterity the evidence that intelligent beings are on earth designing monuments to reveal slowly to the human race. The Intelligence behind the ufo phenomenon knows that the appearance of a superior civilization to an inferior civilization it means the end of the inferior civilization. In fact, we have dozens of examples here in our planet the most known it was the destruction of the Indian cultures in South America by the Spanish Empire. So the Intelligence behind the ufo had been on planet earth since the formation of the Earth and they own the earth and the human race.

The metallic spheres –

The origins of the metallic spheres remains a mystery to archaeologists and historians since they were unearthed in South Africa. Some archaeologist are of the view that this is the handiwork of Primitive men. Although, they discard this possibility because, the formation of the metallic spheres require more intelligence. The scientific analysis it shows that the spheres whoever designed the spheres dating back 2.8 billion years old and were found in mineral deposit. Further, there is no

Evidence that the spheres went through metallic cutting to arrive at the spherical shape. Actually, evolution haven't created Homo sapiens neither mammals or dinosaurs so there was nobody in the planet to create these archaeological findings. Geologists describe the spheres as of a hard surface. Even so, according to geologist Paul V. Heinrich stated that one of the spheres was tested at the California Space Institute and scientists concluded that its balance "Is so fine, it exceeded the limit of their measuring technology "it was within one hundred thousandths of an inch from absolute perfection". Therefore, to conclude this chapter is my conviction that the Intelligence behind the Ufo phenomenon is responsible for these unexplainable archaeological findings. So their

Are showing their presence to the human race.

CHAPTER 2

THE ERRATIC BOULDERS

"We find ourselves faced by powers that are far stronger than we had hitherto assumed, and whose base of operations is at present unknown to us"-Dr. Wernher Von Braun, 1959, on the deflection from orbit Of a U.S satellite. the largest erratic boulder is the "Ayers Rock. "In the Northern territory in Australia, it is the largest single rock in the world. The rock is dark red at sunset it is 21/4 miles long, 11/2 wide with a height of 1.143 feet showing above the ground, it weights 1, 425, 000.000 tones. In fact is called "The heart of Central Australia", because it is composed of red rock, shaped somewhat like a heart and is located at the center, near the geographical center of Australia. As a matter-of-fact, there are no nearby mountains from which it could have been extracted however it is out in the desert with the sand and cactus.

A huge rock for which modern geology can't give an explanation. For this reason, no ordinary iceberg no water currents couldn't transport because a rock 1, 425.000.000 tones is not buoyant. In fact, not even the Antartic ice cap itself or a sizable portion, could lift such enormous weight. Obviously, the ocean have to travel at supersonic speed for the rock not

To sink but that doesn't happen in nature. The Ayers rock is 300 million years old. The rock is so big that it takes 3 to 4 hours to walk around. As a matter-of-fact, geologists believe in slow and gradual sedimentation even the flood is unable to carry such boulders. Moreover, Not even raging torrents at supersonic speed because the rock don't float like the iceberg that displace water. The Ayers rock and big boulder can' t be transported by water or ice. the only way to transport rocks like the Ayers rock is TELEPORTATION by the intelligence behind the ufo phenomenon.

The Gibraltar Rock

The rock of Gibraltar has apparently been turned completely upside down. As a matter of fact, the Eocene rocks are in the lower position, covered with older upper Jurassic which are in turn covered to move

According to geologists the order should be : First Lower Jurassic, Upper Jurassic, with the Eocene on top . is the only way to resolve this confusion is to slide the Jurassic on top of Gibraltar from the east by uplifting the Mediterranean floor which is now at a depth of 2000 meters. The cubic feet of phenomenon use the same technology to teleport the erratic boulders to turn -upside down the Gibraltar rock. The intelligence behind the ufo phenomenon use anti-gravity technology. Furthermore, the loose rocks lying on the Jura mountains were torn apart from the Alps and transported there. In their mineral composition they differ from the rock formations on the Jura, showing their alpine origin. For this reason, rocks that differ from the formation on which they lie are called erratic boulders. The stone blocks that lie on the Jura mountains at an elevation of 2000 feet above lake Geneva, some of them occupied by the are thousands of cubic feet in size.

For instance, The Pierre A Martin boulder is over 10.000 cubic feet and they must have been carried across now occupied by a lake and lifted to the height where they are found. Similarly, there are erratic boulders all over the world. For instance, there are erratic boulders on the shore in the British Islands and in the highlands were transported across the North Sea from the mountains of Norway. Some force wrested them from those massifs and they carry them over the entire space that separate Scandinavia from the British Islands and set them down on the coast and on the hills. In addition, from Scandinavia boulders were also carried to Germany and spread over the country. In some places there were so many as though they had been brought there by masons to build cities. Further, in central Germany, lie stones that originated in Norway. Similarly, from Finland blocks of stone were carried to the Baltic regions flying over Poland and lifted onto the Carpathians. Further, another group of boulders were carried from Finland over the Valdal Hills.

Boulders in North America

Erratic blocks broken from granite of Canada and Labrador were spread over Maine, New Hampshire Vermont, Massachusetts, Connecticut, New York, New Jersey, Michigan, Wisconsin and Ohio. The boulders perch on top of ridges and lie on slopes and deep in the valleys also they lie on the coastal plains and on the White mountains and the Berkshires. In contrast, sometimes they lie in an unbroken chain and in the Poconos mountains they balance precariously on the edge of crests.Indeed, travelers wonder at the size of these rocks and abandoned

There frighteningly up. Certainly some erratic boulders are enormous. For instance, the block near Conway, New Hampshire, is 90 by 40 by 38 feet and weights about 10.000 tones, large as a cargo ship. Moreover, equally large is the Mohegan rock, which towers over the town of Montville in Connecticut ¾ of an acre. Furthermore, the Ototoks erratic boulders 30 miles south of Calgary, Alberta consists of 2 pieces of quartzite originated from at least 50 miles which weights over 18.000 tones formed for blocks of 250 to 300 feet in circumference. In fact, there are small when compared with a mass of chalk stone near Malmo in southern Sweden which is 3 miles long, 1000 feet wide and from one hundred to two hundred feet in thickness and which has been transported for long distance and is quarried for commercial purposes. Similarly, a transported slab of chalk is found on the Eastern coast of England where a village had been built. Similarly, In innumerable places on the surface of the earth, as well on isolated islands in the Atlantic and Pacific, even in Antarctica lie rocks of foreign origin, brought from afar BY SOME GREAT FORCE, broken off from their parent mountain, ridge and coastal cliffs, THEY WERE CARRIED DOWN AND UP HILL, OVER LAND AND SEA. So the Intelligence behind the Ufo phenomenon with technology very advance like anti-gravity machines, teleportation also entanglement.

They exist and operate at the quantum level. For this reason the ufos and its crews can became invisible because they operate at the quantum level. So the Intelligence behind the Ufo phenomenon after creating the chaos with the erratic boulders, the Pleistocene mass extinction and the unexplainable archaeological findings.the Intelligence behind the ufo phenomenon began the creation of monuments around the world. The Pyramids of Egypt, Stonehenge in England, the Platform of Baalbek in Beirut, Lebanon a Gigantic terrace composed of stone blocks, most of them with sides more than 60 feet long and weighing up to 2.000 tones. It is located on a section of the Beirut-Homes railway line and a road at a height of 3, 760 feet. The Platform is incredible old and has no historical date . It is first mentioned in Assyrian writing under the name of Ba'li as early as 804 BC. The Greeks and Romans both made use of it. It is impossible that these stones were transported with wooden rollers or sledges. There is no technical aid in ancient times. no even a crane can lift 2000 tones. Further, the ruins of Sacsayhuaman in Peru. Another archaeological site that defies rational explanation are the ruins of Sacsayhuaman in Cuzco, Peru. The local archaeologists don't believed the Incas built this fortress of stone, situated at a height of 11, 480 to 12, 415 feet. The monolithic block weights more than 100 tones, the ramparts are 18 feet high. The terrace wall is more than 1500 feet long 55 feet high. The rarified atmosphere makes it difficult to breath, the rocks are well polished stones that look as though they were cut out by a cheese knife. Again, there are questions of who created this massive stones?. After all, to have built such immense structures in a terrain so elevated and irregular would have been a fantastic feat that the poor peasants that the poor Indian peasants who populated the area did not have the technology to accomplish. the Stonehenge stones in England, another architectural mystery is the famous stone monument of Stonehenge, which is located about 75 miles south-southwest of London and 8 miles north of Salisbury. It is a popular tourist attraction with nearly 200.000 thousands visitors per year. Stonehenge was built with so called Bluestone, a separated kind of rock with bluish color. Most bluestones are dolerite. The site of the quarry is 240 miles Keep in mind that each stone weights up to 5 tones. Most bluestone are dolerite.

Stonehenge was built between 1900 and 1600 BC, thousands or so years after the pyramids in Egypt. It was a few hundred years before the fall of Troy, and the Minoan civilization was flourishing. Nobody Knows the purpose of the stones or who built them?. Some think they were built for religious or Astronomical purposes. The Druids came after they were built, and the pre-historic people left no evidence of their work. Although, there are many questions,

How were these stones weighing up to 5 tones transported in a very irregular and hilly terrain?

Where was the forest to provide the building wooden rollers?

What kind of tools were used to Quarry the stone?

How were the thousands of workers taken care of?

Further, some believed an enlightened man from Greece had them built. But who was he? Why don't we know the name of the man who accomplished such a feat? Where did pre-historic people get the ideas to build such monuments? Similarly, Europe's persistent drought has caused water levels to plummet. In 2022 submerged monuments resurfaced. The Dolmen of Guadalperal has resurfaced from the depths of the Valdecanas reservoir in western Spain nicknamed the Spanish Stonehenge. The one that just reappeared in Spain dates to the fourth or fifth millennium B.C which makes it as much as 2.000 Years older than its celtic cousin on the Salisbury plain in England. What remains of the Guadalperal complex is a ring of quartzite, 117 feet in diameter, surrounding 144 Jagged granite standing stones, many of which are no longer standing. An atmospheric high- pressure system driven by climate change has left parts of the Iberian Peninsula at their driest in 1200 years. Moreover, the ruins of the Dolmen of Guadalperal are nothing in comparison with the Stonehenge in England. Also the Spanish haven't make a deep study of these ruins. The Pyramids of Egypt moreover, the most enigmatic stone configuration that defy explanation about technique, labor and forms of construction exist around the world. The pyramids of Egypt are a challenge to our modern engineers and technology. The great Pyramid is believed to have been built in 2, 500 BC during the reign of Pharaoh of Cheops. The dimensions are breath taking. It was built with approximately 2, 300.000 stone building blocks weighing from 3 to 70 tones. The base covers 13 acres Which is the equivalent of 8 square blocks in New York City. Napoleon's engineers believed the pyramid of Cheops consisted of enough stones to build a wall around France. The pyramid is big enough to contain all the cathedrals of Florence, Milan and Rome and Still have room for the Empire State Building in New York, Westminster Abbey, St Paul's Cathedral and the English house of Parliament in London. It is calculated that there is more stone in the Pyramid than all the masonry used to build every church in England since the time of Christ. The Cheops pyramid was 485 feet tall. Furthermore all the locomotives in the world could no pull the pyramid because it weights six and half million tones. Some of the individual stones weigh up to 100 tones. The stones are fitted so closely together that is impossible to detect the line where they are joined, How these stones were raised? The Egyptians didn't leave records of the technology used to build the pyramids. The most popular theory is that workers slid the massive blocks up ramps onto the pyramids but there is no evidence. The only written record came from the Greek historian Herodotus, who said it took 20 years and 100.000 men to build the great pyramids. The only problem is that he wrote this 2, 000 years after they were built. Although, there are many questions:

How were the thousands of workers taken care of and fed?

Where did they sleep?

How were they transported?

An army so vast but no Cemeteries are found near the pyramids?

We know that Egypt was not a big holding society. How did they get enough people to hack out 2, 600.000 giant blocks from the quarries? what kind of quarries were they?. what kind of tools did they use? there was no dynamite or other explosives at that time. After the rocks were quarried, how were they dressed? dressing many of these stones would require a minimum pressure of two tones. How primitive man get the technology to apply a pressure of two tones to the rocks? Some historians believed that the Egyptians use ropes to pull those blocks of stone. where did they get the ropes? where the Pharaoh find an architect who could design a building with such a precise measurements? Moreover, historians claimed that 3 to 100 tones were pushed over land on wooden rollers. trees were chopped down, dressed out to logs and used as rollers under the blocks and taken to the building site. The theory is good the only problem is that there are no forests in Egypt. The Great pyramid would have required 26 million wooden rollers. Finally, to import them would have required the largest fleet in history. Where did they get the wood to build such a fleet? In conclusion, only the Intelligence behind the ufo phenomenon had the technology for these superhuman feats with their antigravity, teleportation and quantum entanglement it was easy for them to create these monuments, they were able to create these beautiful structures.

CHAPTER 3

THE UFO EXPLOSION AT THE TUNGUSKA FOREST IN SIBERIA.

> "The disturbing reality is that for none of the thousands of well-documented extinctions in the geologic past, do we have a solid explanation of why the extinction occurred. We have many proposals in specific cases, of course Trilobites died out because of competition from newly evolved fish ; Dinosaurs were too big or too stupid ; the antlers of Irish Elk became too cumbersome. These are all plausible scenarios, bur not matter how plausible, they cannot be shown to be true beyond reasonable doubt. Equally plausible alternatives scenarios can be invented with easy, and none has predictive power in the sense that it can show a priory that a given species or anatomical type was destined to go extinct."
>
> —David M. Raup

The explosion of June 30, 1908 in the Tunguska forest in Siberia was a ufo. In 1921 the new Soviet Academia of Sciences commissioned a remarkable scientist named Leonid Kulik to collect information about the explosion in the Tunguska forest in Siberia. The local newspapers of Irkuts, Tomsk, and Krasnoyarsk had all reported the event. The newspapers described as the most "Unusual phenomenon of nature". In the village of Nizhne -Karelinsk in the northwest high above the horizon, the peasants described"a shinning body, very bright (too bright for the naked eye) with a bluish -white light. It moved vertically downwards for 10 minutes the body was in the form of a pipe (cylindrical). More peasants reported,"The sky was cloudless, except that low down, on the horizon, in the direction in which the glowing body, a small dark cloud was noticed. It was hot dry and when the shinning body approached the ground it seemed to be pulverized and in its place a huge cloud of black smoke was formed and a loud crash, not like a thunder but as if from the fall of large stones, or from gun fire was heard. all the buildings shook and at the same time a forked tongue of flame broke through the cloud". Ufos appear in cylindrical, oval, and many different forms. The peasants wept because they though that the end of the world was approaching.

Furthermore, the village Nizhne-Karelink was situated about 200 miles (320 km) from the centre of the explosion. According to a local meteorologist named Voznesensky. The explosion could be heard 500 miles away from the center (800 km) and at that distance, the seismic instruments in Irkutsk had registered an Earthquake proportions. The peasants reported a "fiery, heavenly body, a flame that cut the sky in two and a pillar of smoke". Another account was by peasant named Ilya Petapovich when she was going to the spring for water. She reported "One day a terrible explosion occurred, the force of which was so great that the forest was flattened to the ground, its roof was carried away by the wind and his reindeer fled in fright. The noise deafened my brother and the shock caused him long illness." Moreover, another peasant named Vasiley Okhchen mentioned how he and his family were asleep when, together with their tent, the whole family went flying into the air."All the family were bruised but Akulina and Ivan lost consciousness. The ground shock an incredible long roaring was heard, everything round about was surrounded in smoke and fog from burning falling trees. Eventually the roar died away, but the forest went on burning. We set off in search of the reindeer, which had runaway. Many did not comeback. "Similarly, another peasant from the village of Vanavara, reported "I saw the sky in the north open to the ground and fire pour out. We thought that stones were falling from the sky and rushed in terror and rushed in terror leaving our pale by the spring. When we reached the house, we saw my father unconscious lying near the barn. The fire was brighter than the sun, during the bangs, the earth and the huts trembled greatly and earth came sprinkling down from the roofs." In addition, there were stories of horses bolting with their plough, and a man felt his shirt burning in his back. MANY WILDLIFE LIKE REINDEER and DOMESTIC DOGS DIED Also many houses were destroyed. The scientist Leonid Kulik, who was sent by the Russian Academy of Science, reported, "Trunks of the falling pines, they all lay with their tops uniformly towards the southeast". He climbed the heights of the ridges and he saw "stretching as far as he could see at least 12-16 Miles (20-25 km) was utter desolation. The huge trees of the taiga, lay flat, firs, pines, deciduous trees all had succumbed. The sharp outlines of the winter landscape etched like a plate and again, this bizarre and unbroken regimentation". Furthermore, despite the destruction the trees still lay in one direction. Whatever had caused the Tunguska explosion, Kulik thought that it had destroyed 37 miles (60 km) in one direction alone ; this was the epicenter. Interestingly, the most careful search of the epicenter and nearby terrain DID NOT PRODUCE A SGN OF A METEORITE IMPACT! Through the aerial survey of 1938 and the close examination of the trees, their burn marks, and the pattern of their fall gave the scientists some idea of what happened during the explosion and the true size of the devastation was approximately 770 square miles. The size was as big as Birmingham in England or Philadelphia in the United States. Indeed, there were odd features in the area of the explosion like right in the middle, a large number of trees were left standing, though stripped of their branches. After all the surveying, digging and searching THERE WERE NO SIGNS OF ANY IMPACT HAD HIT EARTH!!. After the explosion the peasants went to look for they reindeer and they found they charred reindeer carcasses. Similarly, A peasant reported "I was sitting on my porch facing north, when suddenly, to the north -west, there appeared a great flash of light, there was so much heat that my short burned off my back. I saw a huge fireball that covered an enormous part of the sky. Afterward, it became dark and at the same time I felt an explosion that threw me several feet from the porch. I lost consciousness. "As a matter of fact, this witness was 40MILES AWAY FROM THE EPICENTER OF THE EXPLOSION!! In Central Russia, a deafening roar terrified small towns and villages. Obviously, a powerful ballistic wave pushed before the descending object, trees were leveled, Peasant's huts were blown away, and men and animals were scattered like paper. In fact, the explosion was of such a force that the seismographic center at Irkutsk,

550 miles to the south, registered tremors of earthquake proportions. The vibrations traveled 3.000 miles through the ground to Moscow, St Petersburg and the earthquake observatory at Jena in Germany, 3, 240 miles away. Even as far away as Washington DC and Java seismographs were activated by the immense blast. Immediately a gigantic pillar of fire appeared in the sky, and in towns 500 miles away, people heard a series of thunderous claps. The noise was so great that some peasants closer to the explosion were deafened. Others were thrown in a state of dazed shock that made them speechless.

Furthermore, with the brilliant fire in the sky, a searing thermal current swept across the forest, scorching the conifers and igniting fires that continue to burn for days. The heat was felt by people 40 miles away. In addition, at a distance of 375 miles to the southwest, hurricane like gusts rattle windows and doors, collapsing ceilings, shattering windows and flinging people into the air. Similarly, in Kansk, a station town with a newly completed Trans-Siberian railway people walking or on rafts were hurled into the river. Similarly, farther south horses stumbled and fell to the ground. Also passengers in the Trans-Siberian express were frightened by loud noises and were jolted off their seats. The train itself was shacked wildly in the tracks. After, a dark mass of thick clouds rose to an altitude of more than 12 miles above the Tunguska region. Right after, the entire area was showered by an "ominous black rain".

Furthermore, intermittent rumblings of "thunder resembling heavy artillery, reverberate throughout central Russia. Likewise, five hours after the blast, the air waves created by the object, traveled west beyond the North sea causing strong oscillations at meteorological stations in England. Similarly, within 20 minutes sudden fluctuations in atmospheric pressure were detected by newly invented self-recording barographs at six stations between Cambridge, fifty miles north of London, and Petersfield, sixty -five miles south of London. Baffled meteorologists thought that a large atmospheric disturbance had occurred somewhere in the world. Actually, after two decades, scientists discovered that their 1908 barographic records correspond with the cataclysmic explosion of June 30, 1908 in the Siberian Tunguska forest. Moreover, the airwaves circled the earth twice. Also, astonishing was the effect on the earth magnetic fields.

In the 1960's a scientist, D.R. Vasilieyen pointed out "The evidence of electromagnetic chaos at the center of the explosion" also he said there was an "electromagnetic hurricane of enormous proportions, which has shattered, perhaps permanently, all the normal alignments with the earth's magnetic field". Furthermore, Russian researchers noticed the similarity between the destruction at Hiroshima and the Tunguska explosion because there had been at least two blast waves, extensive fires and flash burns. also, the scientists noted an accelerated growth of trees and plants.

The first American observers noticed that right in the center of the explosion, it had little damage and some trees remained remained upright similar to the nuclear explosion at Hiroshima. Also, at Hiroshima the trees seemed to grow rapidly. The similarities between the Tunguska explosion and Hiroshima are too close to ignore. Although is beyond belief to think that an atomic explosion had occurred in Siberia. The first atomic explosion was carried out by the United States forty years after at Alma Gordo. Dr. Vasilieyen of Tomsk University said, "there had been the most violent genetic changes, not only in plants but in small insects life. There are ants and other insects quite unlike anywhere else. Some of the trees and plants just stopped growing, others have grown many times faster than before 1908". By the same token, Scientists found more similarities between the nuclear tests carried out by the British, Americans, and Russians and the Tunguska explosion in Siberia. Moreover, The Tunguska explosion extraordinary bright aurora lights and disturbances in the Ionosphere. Dr. Vasilieyen stated," It is certainly odd, I know of no other phenomena than the nuclear explosions, which produce these displays at their magnetic opposite side 0f the world, though it could be just coincidence". Another important feature of this phenomena is that like in Hiroshima, the object that exploded at the Tunguska exploded In the air. Likewise, after the explosion, a mass of silvery clouds appeared in Russia and Northern Europe. The light was so intense that during the next few nights people took photographs at midnight and ships could be seen for miles out of sea. A Russian scientist described the phenomenon as "A thick layer of glowing clouds as lit by some kind of yellowish green light that sometimes changed to a rosy hue. It was first time. I had seen such phenomenon."

Additionally, extraordinary dust clouds and eerie nocturnal displays are observed for weeks across the continents as far south as Spain. Interestingly, on June 30 a scientist in Holland reported, "An undulating mass passing across the northwest horizon, it was not cloud for the blue sky itself seemed to undulate". By the same token, after sunset in Antwerp the northern horizon appeared to be on fire". Conversely, in 1930, at the Royal Meteorological Society Quaterly journal, a scientist gives an account of the odd colors he observed over England in 1908 on the nights of June 30 and July 1 1908 "A strong orange -yellow light became visible in the North North-East causing an undue prolongation of twilight lasting to day break and there was no real darkness". The phenomenon was reported from various places, in the United of Kingdom, Copenhagen, Konigsberg, Berlin and Viena. According to the London Times of July 4, 1908"The remarkable ruddy glows, which have been seen over an area extending over an area as far as Berlin". Witnesses reported that abnormal glows appear in the sky only after the fading of twilight. The sky grows partially dark and then brightens again with lurid color". A witness in London reported "It was possible to read large prints and the hands of the clock and the hands of the clock in my room were distinct ""An hour later at 1:30 am the room was quite light as it had been day". Another witness reported, "The northern sky t midnight became light blue, as if down were breaking and the clouds were touched with pink in so marked a fashion that police headquarters was run up by people who believed that a big fire was raging in the north of London". In the London suburbs, people were drawn into the street to view the frightening cosmic phenomenon. A meteorologist said "I have never at night time seen anything the least like this in England and it would be interesting if anyone would explain the cause of such unusual sight".

Furthermore, the similarity between the Hiroshima event and the Tunguska explosion, In the Hiroshima event wood was ignited at a distance of one mile. On the other hand, in the Tunguska explosion Wood ignited at 8 to 10 miles. Similarly, from the fall point in Hiroshima the blast destroyed an area of approximately 18 square miles. At the Tunguska explosion, the area destroyed was 1200 miles. At the Tunguska explosion people felt the heat 315 miles from the blast. On the other hand, In Hiroshima people felt the heat 60 miles from the fall point. In addition, Russian and American thought in the possibility that the Tunguska explosion was"100 times more massive in the megaton range. "The Nobel prize winner chemist Willard F. Libby has estimated its energy yield as high as 30 megatons or 1500 times as great as that at Hiroshima. Moreover, In 1966 A report Russian scientists determined that the Barometric and magnetic effects as well as the zone of destruction of the Siberian forest were similar with the high altitude Nuclear tests conducted by the United States. Interestingly, if the Tunguska explosion had taken place in 1958 the advance instruments at the observatories would have detected a high altitude atomic explosion had occurred. Likewise, some scientists were able to calculate the height at which the object had exploded in the air at the Tunguska explosion. Scientists though that the object had exploded at 5 miles (8 km) up in the air. But what could have caused the appearance of a nuclear blast before the birth of the atomic age? Also scientists though that the object weighed hundreds of tones.

Witness accounts of the object that exploded at the Tunguska forest a newspaper published in Irkutsk, a village situated 550 miles from the explosion, reported that on the morning on June 30, 1908 in a village north of Kirensk, peasants describe a body "shining very brightly. The villagers had dashed out into the streets in absolute panic; some wept in terror convinced that this must be the end of the world ".The peasants stated, "and the pipe shape Did not sound like a normal meteoric object, the cloud of black smoke and the flame were also baffling." Certainly, reports of an enormous fiery object had been seen over villages and towns throughout the Yenisei river province. Some peasants described the object

Like almost moving horizontally from the south, and everybody heard and felt tremors and loud explosions. Peasants reported, "A subterranean crash and roar as from distant firing, doors, windows and lamps were all shaken. Five to seven minutes later, a second crash followed louder than the first, accompanied by a similar roar and followed after a brief interval by yet another crash". Leonid Kulik, the scientist send by the Russian Academia of Science thought that it must have been a giant asteroid, and as a result, a giant crater must have been seen in the ground. The comet and meteorite found no explanation for the seismic shock that registered around the world on June 30, 1908. Science had never noted shocks or earthquakes like tremors in the case of an asteroid striking the earth. Also, never meteors or asteroids had disturbed earth' magnetic field or gravitational field. Scientific research in the 1960's failed to find any electromagnetic disturbances caused by meteorites approaching the magnitude of the Tunguska explosion of June 30, 1908 in the Siberian forest. Interestingly, even if the technology for earthquakes was just beginning to be created. The primitive instruments Of the day were able to register worldwide tremors. According to newspapers reports, the object had been observed over an area of 500 miles. From a Krasnoyarsk newspaper of 1908, which reported that on several villages along the Angara River, in the center of the taiga peasants saw a "Heavenly body of fiery appearance cut across the sky from south to north. When the flying object touched the horizon, a huge flame shot up that cut the sky in two. The glow was so strong that reflected in widows faced north". Furthermore, in other villages horses and cows began to whinny and run wildly. I think probably during the Pleistocene mass extinction the animals behave like that.

Likewise, the peasants stated, "One had the impression that the earth was just to gape open and every

Thing would be swallowed in the abyss". Similarly, in the Trans-Siberian tailway the station agent reported, "that he had felt a strong vibration in the air and heard a loud rumbling sound". The locomotive engineer had become so frightened by the ground tremors and noise that he halted his train, fearing it might derail. As a matter of fact, investigators had found no signs of a meteorite or asteroid, peasants mention a "A huge area of flattened forest". Another witness reported "An unbelievable loud and continuous thunder; the ground shook burning trees fell and all around there was smoke. Soon the thunder stopped, the wind ceased but the forest continued to burn". In addition, peasants reported" The reindeer had scabs that never appeared before the fire came. "Obviously, what the peasants reported was a ufo, their descriptions is similar to other descriptions given out throughout history. For instance, in Russia on August 15, 1663 peasants reported that between 10 am and noon, "A great noise resound over Robozerd lake ; from the north out of a clear sky appeared A HUGE FLAMING SPHERE no less 130 feet in diameter emitted two flames beams about seven feet in length ;from its sides poured bluish smoke this huge ball of fire hovered over the lake, after the great flame and two smaller ones vanished". The phenomenon was observed by dozen of witness, and some peasants were on boats in the lake, but the scorching heat force them out of the water. Also the peasants reported that the light from the object, had penetrated the water and reached the bottom of the lake and the fish fleeing from the flame toward the shore". Furthermore, one of the oldest reports in Russia on ufos goes back to the year 1028 . In the Russian Chronicles there is a report of a sighting

Of a serpent -like object that could be seen all throughout Russia. The Object hovered for two days in a fiery pillar. IT APPEARED FROM THE GROUND, ACCOMPANIED BY thunderous noise . The witness thought that it was a sign from god. Likewise, in 1317 over the city of Tuer a ufo was seen for over a week. Similarly, on April 9, 1628 in Berkshire, England a witness reported

"The weather was warm and suddenly a hideous rumbling was heard in the air followed by a strange and fearful pearl-like object and thunder. It sounded like a rough battle a great cannon seemed to roar then the second time two cannons shots seemed to have been discharged in the sky. then there was a sound heard like the beat of adrum, sounding a retreat, then a hissing sound." Similarly, On December 5, 1735 at 5 pm in England a witness reported the appearance of a "deep red cloud under which a luminous body sent streams of very bright light by which the witness could easily read. Shockingly the streams of light moved slowly for some time and stood still!!". The witness proceed that" it became so hot that I had to strip my shirt in the open air it looked like a great ball of fire". Likewise, on December 11, 1741 a witness reported the sighting of a ball of fire. A most terrible clap of thunder was then heard in the North, like two very large cannons fired a second after each other, but the rolling and the echoing were not like cannon shots. All the houses were shaken 20 miles around". Moreover, in the 20 century one of the most important ufo sightings, the object reported look just like the Tunguska object. On June, 4, 1965 the American astronaut James Mcdivitt during the flight of Gemini IV flight saw a cylindrical object ahead of their spacecraft. After the flight Mcdivitt checked the tecords and there were no rockets near the Gemini capsule at the time of the sighting. Similarly, Gordon Cooper and Charles Conrad, during the flight aboard Gemini V on the third day of their mission, as they were passing cape Canaveral (at that time it was called Cape Kennedy) The Gemini flight director, Christopher Kraft, asked them" if they could see anything flying alongside the Gemini". The director radioed to the astronauts," We have a radar image of a space object going right along with you. From 2000 to 10.000 yards away." The object radar returned it was approximately the same magnitude as Gemini v. Interestingly, the object was tracked until both Gemini and the object were lost beyond the curvature of the earth. here are the similarities between the Tunguska object and ufos.

UFOS

1. Witness reported thunderclaps
2. Some ufos look like tubes and they can change form into a oval or a plane form. James Mcdivitt in the Gemini V witness a a tube or cylindrical ufo during their Flight on August 24, 1965
3. Ufos look like flaming objects
4. Ufos appeared and disappeared in a blink of the eye
5. Ufos produce heat, radiation, electromagnetism and earthquake like tremors.

The Tunguska object

1. The object had a form of pipe, tube or huge cylindric
2. The object glowed bluish and white radiance
3. The tube or cylindrical object moved slowly
4. The object disappeared in a blink of the eye
5. The object produced earthquakes, radiation disturbances, electromagnetic heat asteroids when impact earth create craters for instance :

Name - location - Diameter - Age
Chiexulub - Yucatan -Mexico 180 km - 65 million years (caused the extinction of the Dinosaurs)
Chesapeake Bay – Virginia -US 85 km - 35 milion years old

6. Kara - Russia 65 km Diameter - 57 million years old
7. Popigal - Siberia -Russia - 100 km - 35 million years old
8. Saint Martin -Manitoba -Canada - 40 km - 220 million years old
9. Acraman – South Australia - 85-90 km - 580 million years old

So, is clear that asteroids and big meteorites leave a crater when they impact the earth.

However, in the case of the Tunguska explosion in the Siberian forest on June 30, 1908.

We see that the explosion was in the air. The Flying Path of the Tunguska ufo according to Russian Scientists the 1908 flying path of the ufo and probably location of the blast had been estimated in the Mid-1920s by a Russian scientist A.V. Voznesensky, former head of the Irkutsk observatory. The Scientist used some information bay Kulik and Obruchen and earlier seismic data from Iskutsk and other Russian stations and observations of acoustical phenomena throughout Central Siberia. Voznesensky attempted to trace the path of the object and find the location of impact. He found out that the effects of the explosion" had been heard by people over an incredible immense geographical area. One larger than France and Germany combined. "The fiery object racing through the cloudless sky, had been observed by thousands from the southern border of Siberia to the Tunguska region. Similarly, the noise of the explosion the thunderclaps and rumblings like thunder was heard for a radius of 500 miles. From the reports and seismic data, he was able to figure out the time of the blast at about 7:17 AM on June 30, 1908. The place of the fall, he though was in the territory North of Vanavara in Siberia. Moreover, in February, 1927 scientists departed from Leningrad with a research team to Kansk and farther east to the remote station of Taishet, and they were agree on the northward direction of the object, and the thunderclaps heard 500 miles from the explosion. In addition, the scientists during their investigation, found," The oprooted trees showed their tops always pointing toward the south. The direction in which they have been heaved by the blast. The ground was littered with fallen dead trunks, their roots stripped away. The dead trees bore traces of a continuous burn from above. Even the broken limbs of those trees still upright were charred at the break". The most astonishing statement from the scientist is the fact that every broken branch showed signs of fire, indicating that the burns were not of a forest fire but the result of a SUDDEN INSTANTANEOUS SCORCHING. A FLASH OF INTENSE HEAT seared and charred everything. As far as the eye could see stretched enormous dark patches of scorched flattened forest, tress uprooted and lay down at the same angle at the same angle, their tops facing south or southeast.

Moreover, hey reported that "first it seemed like the flying object had entered earth's atmosphere and became visible somewhere lake Baikal and then travelled from south-east to north-east as it plunged downwards though it was some suggestion that it might have changed direction". Interestingly there were more than 700 eyewitness accounts. The Russian scientists believed that the object was a spaceship. Furthermore, soviet academician Zolotov said that the object had been travelling in "Cosmic terms extremely slowly before it exploded. Perhaps as little as one km per second!". Some eyewitnesses suggested that the cylindrical object changed course. Also the irregular shape of the ground damage was like "an outspread eagle's wings". The point of explosion should have been more circular like an asteroid shape. Further, the soviet scientists thought that the explosion took form in some kind of container. The soviet scientists couldn't imagine that the mothership carry with her smaller fleet of ufos that can carry explosives that caused the explosion at the Tunguska Siberian forest. As I write there are videos of ufos dumping something on the earth. Moreover, the object that exploded at the Tunguska forest, it seems was sighted at the final phase of its trajectory before exploded. By the same token, in 1946 named Alexsandr Kazantsev postulate the theory that the explosion at the Tunguska forest was caused by a spaceship from another planet. He stated "The explosion wave rushed downward and the trees directly below the point of the explosion remained standing, having lost only their crowns and branches. The wave burned the points of those breaks on the trees and with the permafrost, spitting underground waters, responding to the tremendous pressure of the blow, gushed up as those fountains seen by natives after the explosion. But where the explosion wave struck at an angle, trees were felled in a fan-like pattern. At the moment of the explosion, the temperature rose to tens of millions of degrees. Elements even those not involved in the explosion directly were vaporized and in part carried into the upper strata of the atmosphere were continuing their radioactive disintegration, that caused luminescent air. In part these fell to the ground as precipitation with radiactive effects". For instance, a tungus tribesman recalled," My Father went into the fallen taiga and saw a huge column of water flowing out of the ground. A few days later, he died in terrible pain as he was on fire. But there was no trace of fire anywhere in his body". Likewise, numerous eyewitness who saw the object before it exploded in the sky, reported that it was of considerable size and was a considerable size and was a cylindrical shape. Also, the witness describe the object in the form of a chimney. .Some witness reported,"The elongated flaming object glowed with a bluish-white radiance, brighter than the sun and left a trail of multicolored smoke in the atmosphere". According to a Russian scientist", In its descent over the Tunguska region the object created a huge ballistic wave, that was exactly the same as the air wave of a missile". The velocity of this cylindrical missile like object was first thought to be as much as 30 to 40 Miles per second. In order to account for the higher kinetic energy of the blast however, the Russian Geophysicist made more accurate calculations of the speed of the object. He based his calculations from comparisons of the effect of the ballistic wave and blast force on trees in areas of the explosion. He calculated that shortly before the explosion"The velocity was probably no more than 1 ½ to 2 miles per second or about 700 miles per hour". Another Russian Scientist professor Ziggel, points out that eyewitness saw the object and heard the deafening roar simultaneously.

Moreover, the object left a streaming trail, spaced explosions the oddly shaped pattern of levelled trees also the sudden growth of vegetation after the Explosion. In 1961, Russian scientists showed the peculiar oval or elliptical shape of the Tunguska blast. A pattern that puzzled the Russian scientists was a trajectory from east to south-east. A Russian scientist E.L. Krinov, during the third Tunguska expedition in 1929, observed "The area of uprooted forest has an Oval with the major axis from situated in a direction southeast to northwest". The oval shape surprised the Russian scientist because they expected a circular shape in accord with the crash of a meteorite or asteroid. Certainly, the 1938 aerial photographs further verified the Oval pattern at the site of the blast. Similarly, R.P Florensky's expeditions in 1958, 1961, and 1962 determined from extensive ground and air survey they confirmed that 1200 square miles leveled by the explosion and the scattering of cosmic dust from the blast had a definitive Elliptical contour. Moreover, maps drawn by Florensky's Team show clearly that" the center of the destruction, the complete scorched area contained the dead but upright trees, lay in an off -center position in an explosive wave that fanned chiefly toward The south and northeast". The Russian scientist wrote about the significance of the odd elliptical Contour of the blast. they wrote" It is very evident on the map of the region that the boundary of the area of complete levelling of the forest is irregular in outline. Also the epicenter of the explosion and the zone of trees left standing, occupy an eccentric position in the region of the catastrophe. Obviously this asymmetry cannot be explained by the effect of the ballistic wave, due to the flight of the body the zone of destruction is elongated in a direction that is not parallel to the trajectory but at a large angle to it". In fact, they characterize the blast as directive because the effect of the explosion was not the same in all directions". Following two expeditions one in 1959 another in 1960 in which all the evidence about the Tunguska explosion to experts found acceptable. He said the"blast had an unusual oval shape, because the explosive material was encase in some type of container. The structure of the container, like the thick paper cylinder of a large firecracker caused two explosive charge, to fan out elliptically as it burst. The directivity of the explosion was due to the inhomogeneity of the container". According to the Russian scientists the object that exploded at the Tunguska forest consisted of at least two parts"(1) A substance capable of a nuclear explosion (2) A non-explosive shell is there any evidence of a nonexplosive container? Indeed, some Russian scientists believed that at least partial proof was found by Kulik in his first expedition. Next, the Russian scientist A.Y. Manotstov'kov made some calculations that agree with Zolotov's findings that the object must have arrived at a velocity much slower, than a natural cosmic body. The object entry speed is comparable to the velocity of a high -altitude reconnaissance plane. Another Russian scientist a rocket specialist examined this evidence and concluded with the other scientist that the object in its entry and velocity behaved like a supersonic craft. What flight path did this craft follow to its trajectory are the reports and observations of eyewitness and the ballistic shock damage caused by the rapid compression of the air ahead of the moving object. The deafening thunder heard in June, 1908 heard by hundreds of people throughout Central Siberia during the flight of the object was probably was caused by its powerful ballistic waves. Similarly the powerful thunderclaps heard resulted from the massive blast waves. According to Russian experts, the object created strong ballistic wave during its flight. Furthermore, Russian scientists studying the different investigations since the 1920's to the 1960's have arrived at very shocking conclusions about the fight path of the object that exploded at the Tunguska forest. Three of the first researchers of the Tunguska explosion Voznesensky, Suslov and Astapovich after collecting eyewitness reports and seismic data stated, that the object moved from south-southeast to north-northeast. According to Florensky's findings "Both the general pattern of the toppled trees and the relationship between of fallen deadwood and the searing effects. As well the distribution of Cosmic dust indicated that the object came from east-south-east to west -northwest was the most probably trajectory". On the other hand, another scientist named Zolotov examined trees that bore traces of the ballistic and blast shock. He concluded that the air wave which caused minor damage compared with the explosion, had definitely come from he soth-west.He added "The object had been visible overhead as a "Fiery object" to villages near kansk southwest of the blast but it had also been seen in KIrensk and other towns lying to the southeast. According to the scientists, dozens of reliable observations made both flights paths equally possible. Some scientist disagree" The same object could not have appeared almost simultaneously in two different locations hundreds of miles apart or could it." Ultimately, the scientist solved the problem of the trajectory with a shocking answer "Bothe paths were accurate; the object had switched direction in its journey over Siberia. Furthermore, the research by Florensky and Zolotev about the ballistic shock effect on the trees provides a strong basis for a reconstruction of an alteration in the object's line of flight. According to the scientist, the object in the terminal phase of its descent the object appeared to have approached on an eastward course, then changed course westward over the region before exploding". The ballistic wave evidence in fact indicates that some type of flight correction was performed in the atmosphere. "The Russian scientist Felix Zigel postulate the same theory . His study of all the eyewitness and physical data convince him and all the scientists researching the phenomenon that the object prior to exploding changed from an eastward to a westward direction over the Tunguska forest. Prof. Zigel added, "Before the blast, the Tunguska object describe in the atmosphere a tremendous Arc of 375 miles in extent in azimuth that carried out that it carried out a maneuver. No natural object is capable of such a feat!! Moreover, The Russian scientists and roquet and aviation experts Manotzkov, Liapunov, Kazantsev and Zigel agreeing that the"Cylindrical object causing an elliptically -shaped atomic blast in 1908 could only have been an artificial flying craft from another planet. In addition, to its maneuvers near the earth's surface, the craft must have steered, as it approached from outer space, into a trajectory angle almost identical to the re-entry path used by modern spaceships. Isaac Asimov gave his opinion about the Tunguska explosion "A fall like that in the middle of Manhattan would probably knock down every building on the Island and large numbers across the rivers on either side, killing several million within minutes of impact". Furthermore, all the Evidence of the June 30, 1908 explosion in the Tunguska forest in Siberia has been photographed, measured and analyzed. Scientists agree that a spaceship caused an explosion of approximately 30 megatons. The scientists ask how could an atomic blast occur in an age when the human race had no nuclear technology. The seismographic center at Irkutsk at 650 miles to the south registered tremors. In Jena, Germany tremors are registered 3,240 miles from the epicenter. Similarly, as far as Washington tremors were felt. The explosion at the Tunguska Forest in Siberia it was bigger than the following:

1. The 1500 BC. Explosion in the Island of Santorini in the Aegean sea
2. The eruption of the Vesuvius
3. The eruption of the Krakatoa
4. Bigger than the Hiroshima bomb
5. Bigger than theb1950's nuclear test

The nuclear blast at the Tunguska forest in Siberia was comparable to 10 to the 23 ERGS Would be to the explosion of the heaviest Hydrogen bombs. However, for numerous reasons it was a silence in the scientific world from June 30, 1908 to 1921 is understandable it was beyond human understanding. Furthermore, The Intelligence behind the Ufo phenomenon with their super-advanced technology they can cause the following events :

1. The removal of magnetism of earth's magnetic field
2. They can affect the change of weather from tolerable to extreme. They can increase or disminution of ice at the polar caps.
3. they can affect the drift of the crust
4. They can affect the volcanoes. Ufos have been seen inside active volcanoes. They are videos.
5. They can create earthquakes
6. They can manipulate the DNA and create new diseases.

The Intelligence behind the Ufo phenomenon exist and operates at the quantum level.

For this reason they can do many things that are impossible for human beings,

Everything is about technology ; Even the spiritual world is about technology.

Chapter 4

The Mystery of the Fossil Graveyards.

> *"The Pleistocene period ended in death. This was not ordinary extinction of a vague geological period, which fizzled to an uncertain end. This death was catastrophic and all inclusive. The large animals that had given the name to the period became extinct. Their death marked the end of an era. But how did they die?. What caused the extinction of 40 million animals?. The extinction was of such colossal proportions as to be staggering to contemplate the corpus delict may be found almost anywhere the animals of the period wandered into every corner of the new world not actually covered by the Ice sheets, their bones lie bleaching in the sands of Florida and in the gravels of New Jersey".*

> —Professor. Frank C. Hibben

H Verril a paleontologist wrote : "In an arid valley in Chihuahua, Mexico paleontologists found countless thousands of skeletons of giant Bison, Mastodons, Camels Six Horned Antelopes, Horses, Rabbits and gophers besides carnivorous like wolves, Sabre -toothed tigers, hyenas, badgers and bears. "His theory is that the Glaciers from Canada moved southward and in its way dragged" vast herds of Bison, Mammoths, Deer, horses and Mastodons". South into Texas and into the valleys and plains of Mexico. They perished in this desert valley were paleontologists found their bones. So paleontologists invent theories to try to explain the thousands of bones found like the" starvation theory "that claim the animals eat all the grass and they died of starvation where they were found. Another theory is that a "sudden movement of the sea". Caused a supergiant tidal wave. This theory was proposed by Alcide D'Orbigny, one of the successors of Charles Darwin in order to explain the thousands of skeletmpianons of horses, mastodons, Armadillos, the glyptodon and megatherium, buried in the great Pampian, Argentinian in South America. He claimed "I argue that this destruction was caused by an invasion of the Continent by water". The immense number of bones were carried out to the coast. The sudden movement of the sea which invade all at once the continent according to some scientists was the result of the sudden rise of the cordillera of the Andes. The sudden rise of the Andes is said to have shaken the whole world with a mighty earthquake According to R.T Chamberlin and James Dana" Hundreds if not thousands of cubic miles of the Earth almost instantaneously raised upward produced violent earthquakes which spread throughout the entire world. The Rockies and the Andes forming a chain" A third of the circumference of the Globe had simultaneous orogenic movements with the sudden movement of the sea invading the continents. Furthermore, there is a ridiculous theory name" The ancestral Cemetery Theory". This theory never has been accepted as rational or with common sense. The theory try to explain the 20 tones of Hippopotamus bones dug out of the top of a mountain they travel there from Africa to die there. So when each animal felt death approaching it would travel to Palermo, Sicily to the hill and die with their ancestral relatives. The same way human beings are buried in family plots. Actually, Sir Joseph Prestwick of Oxford argued against "The ancestral cemetery theory". That was current in his day pointing out that the bones" Are those of all ages down to the fetus". Obviously, is hard to think that baby hippos following their parents all the way from Africa and died in the hills of Palermo, Sicily. So vast numbers of hippopotami were swept into caves overtaken by waters. The theory the Ruthless Men slaughtered these animals in a cave have been dug out of one hill, bones of rhinoceroses, Camels, giant wild boars, bears, horses, deer wolfs etc. What could have killed so many

Animals all at one spot? a tornado, a pestilence, an earthquake, a tsunami, the intelligence behind the ufo phenomenon? In fact, it has been dug out "Sea shells, plants and trees all mixed like a salad bar large, small, herbivorous, carnivorous, mammals and birds. All mixed in great confusion all in one pile in alluvial deposits. Also, is believed that primitive man set fire to the prairie grass to stampede the Mammoths over the cliff. Moreover, the accession of the mammals to the Dinosaurs has never been explained nor the dying of the mammals in the Pleistocene. According, to Charles Darwin," All attempted explanation have failed to solve the mystery of extinctions and fossil grave yards". As a matter of fact, Sir Henry Howorth writes," Darwin who have looked the problem face to face, confessed to me that it remained to him the one stupendous mystery in the in the latter geological history of the world for which NO RATIONAL EXPLANATION HAD BEEN FORTHCOMING". Similarly, a century after Darwin, geologist James H. Zuberge of the University of Michigan admits". We are still unable to account for the loss of the gigantic animals of the Pleistocene". By the same token, a few geologists are willing to admit quite honest that the RIDDLE OF THE MAMMOTH and the other millions of animals that disappear in the Pleistocene is an INSOLUBLE CONUNDRUM.

The Spring Flood Theory

In the Artic ocean off the coast of Siberia are the new Siberian Islands are full of remains of Elephants, Mammoths and other Pleistocene mammals. The entire islands seem to be composed of "organic debris". The most numerous are the mammoths, elephants and fossil trees and Plants. A total of 66 animal species in these new Siberian Islands including reindeer, artic fox, horse, tiger, bear, musk, ox, antelope and woolly rhinoceros. Since the Universal flood theory was not accepted in the 1830s.Geologists have been unable to present a rational explanation to the mass graveyards. Next, The fossil graveyard at La Brea, in Los Angeles, California I believed "La Brea " fossil graveyard is the most famous in the world. The Brea asphalt pits are situated inside the city limits of Los Angeles. The primary theory is that herbivorous, elephants, camels, horses mammoths, deer, antelopes, got stuck in the oily quagmire. Obviously, the carnivorous, tigers, hyenas, saber tooth tiger, leopards, bears followed them and jumped to eat the struggling victims. Scientists have discover 3,000.000 bones from La Brea Asphalt Pits. They discover 700 Sabre tooth tigers, from Pit stuck number 3 discover 17 Imperial Elephants, from pit number 9 camel, horses, ground sloths, mastodons, oak trees, cypress trees, birds larger than a condor, coyotes, mountain lions, human being. The theory is that as the centuries went by and animals from nearby hills and mountains walked by got stuck permanently followed by flesh -eating animals. So, thousands of animals got stuck and died struggling in the Tar. All these animals were swept down the mountains sinking slowly in the quagmire.

The Origin of the La Brea Asphalt Pits

The Intelligence behind the Ufo Phenomenon is the creator of the Tar Pits discovered in many parts

Of the world. Certainly, thousands of legends mention Ufos but with different names in accord with the social stage in the development of society. Moreover, thousands of legends, traditions, and Myths mention a calamity that took place within recent human history. For instance, the Greeks Legends call a visitor that caused destruction in the world Phaeton. actually, what were describing were ufos but they didn't have the technological terms to describe them because they live in a pre-industrial society and they couldn't describe spaceships. So, they thought that were seen comets, moons, planets creating wonders in the heavens. The traditions and legends mention STICKY and INFLAMABLE BLOOD LIKE FLUIDS FALLING FROM THE SKY as Hot NAPHTHA which is a colorless flammable liquid obtained from crude petroleum, used as and a raw material for gasoline. Bitumen which is any of various mixtures of hydrocarbons and other substances occurring naturally or obtained from coal or petroleum found in asphalt and TAR. Furthermore, ancient legends and traditions mention rains of fire which are of hydrocarbons origin. Also fluids described as bloodlike were not blood but of a brownish red liquid. For instance in a legend in Finland, it is called red milk; this implies that it was thick and opaque like the sticky La Brea Tar Pits. Likewise, Greek legends of Phaeton (an assumed star) mentions that rains were hail mixed with blood ; in fact it was ice mixed with liquid ferruginous and metalliferous matter. By the same token, the bitter taste of the waters caused by Phaeton were obviously chemical reactions between incompatible substance. Interestingly, a great concentration of ferrous minerals and traces of ferruginous staining exist in various terrestrial surface deposits, obviously laid down by ufos or like the Greeks and Romas called it Phaeton.

Furthermore, bones of late Pleistocene animals have been found in drift deposits to be stained with iron. Additionally, fossil remains found at hi terrace gravel at Acton and Turham Green, West London, were found with abundant traces of Manganese deposits a Brittle-Gray metallic element loaded with Manganese -blue gray -iron sands. However, the most perplexing conundrum is why thousands of animals humped to their deaths? into the Tar Pits. We know that humans and animals have an strong instinct and will to live and survive. We can see that every day, for instance, in the street a herd of ducks before crossing the street wait for the traffic to stop and then they cross the street. Obviously, the will to live and survive. so they don't have to be rocket scientists is natural to want to live in animals and humans.

Lately, in many parts of the worlds people are witnessing animals going around and around in what is called "Circles of Death". So, the animals in the Pleistocene Mass Extinction were ordered to jump by Intelligence behind the Ufo Phenomenon in the Brea Tar Pits . My theory is that the Intelligence behind the Ufo Phenomenon used a telepathy -magnetic sense -electroreceptor metathetic stimulus. Biologists and naturalists know that animals in general have the capacity to detect the Earth magnetic fields, which acts as an aid for navigation and in their yearly emigrations. For instance, the navigation of European Robins can be altered by superimposing an artificial magnetic field. Similarly, pigeons with magnets attached to their heads became disoriented. The same idea but with more advanced technology the intelligence behind the ufo phenomenon made the animals in the Pleistocene to jump into the Tar Pits. So, how this "jumping "theory can explain the Mountain Lion and his cubs, the Tiger and his cubs, the wolf and his cubs dug out from the asphalt, showing that the cubs were in development, the skeleton and dentition. We all know that the tiger, wolf and mountain lion nurse their babies in a cave or Den and certainly, the oak and cypress trees didn't jump in after each other. Similarly, there is a Tar Pit full of fossils in Ecuador and two smaller tar pit in California. After all geologists cannot explain how these pits got started operating in the first place. In fact, asphalt is an organic matter. So, from were the tar come from?. The answer is the Intelligence behind the ufo phenomenon created the mystery of the Cumberland Bone Cave.

The Cumberland Bone Cave is situated in Maryland where hundreds of different species of animals found refuge and they die in the cave. In this one cave Have been found wolverine, grizzly bear and Mustelidae which are native to colder regions. Moreover, they found peccaries, tapirs and Antelope related to the present day Eland. Also, animals indigenous to tropical regions like ground hogs, rabits, coyotes and hare indicate dry prairies. Further, animals of water habitats as Beaver and Mushrat. Nevertheless, what made rabbits run into the same cave as coyotes? and Antelopes run with wolverine, grizzly bears? Also bones of the Mastodon were found and reptiles. Most of the animals were mammals. J.W. Gidley a scientist that investigate the cave declared that the bones were so "Hopelessly intermingled " that the only sensible conclusion was that the animals were contemporaneous and were deposited at one time. This refutes the theory that the Northern species were deposited during the Glacial period and the Southern species during the interglacial period animals of Northern regions like :wolverine, lemming, the long -tailed shrew, mink, red squirrel, muskrat, porcupine, hare and elk scrambled with animals of warmer climates like peccaries, crocodiled and tapir. Also, animals of arid regions jumbled together like coyote, badger, horses, Death came to all of them at the same time. How can we explain the different species of mammals of different geographical regions intermix together?

Nature don't work this way it can't choose and pick the different species of animals and jumbled together in a cave. Nature don't have self-consciousness and choose and pick the different animal species. Only the Intelligence behins the Ufo phenomenon can do this with a technology name "Quantum Entanglament"; They can operate all over the earth. Obviously, the ice, floods, mud and water don't have the property of mixing different fauna and flora of different geographical areas.

The Mediterranean fossil Graves

A great scholar, Dr. Hugh Falconer explored the Sicilian cave of Moccagnore Stated "The cavern had been filled right to the roof, the uppermost layer consisting of a concrete of shells, bone splinters, with burn clay, flint chips, bits of charcoal and hyena coprolites, which was cemented to the roof by stalagmitic infiltration of contemporaneous origin, occurred to the conditions previously existing, emptying out the whole base, incoherent contents and leaving only portions AGGLUTINATED TO THE ROOF". Moreover, in the hills of Palermo, Sicily thousands of Hippopotamus bones have been found. The bones were of animals of all ages, even a fetus, and they show traces of weathering or exposure. The scientists researching this cave reached the conclusion, that the freshness of these bones and the presence of very young animals that some

"Sudden and geologically recent cataclysm is responsible for the accumulation of bones". After all we don't know any agency in nature that can cause a cataclysm of these magnitude. Scientists found a great variety of animals. For instance, they found hyenas, lions, bears, tusked elephants, a dwarf hippopotamus. The animals were found in caves and rocks often TOO SMALL even for one animal. Also, most of these animals did not live where they were found. In fact, It seems that THEY WERE TRANSPORTED from very DISTANT REGIONS. Moreover, remains of turtles and tortoise were found in FISSURES. Likewise, Because of the freshness of the bones, it seems that many animals have been entombed in the flesh and in dismembered condition. Shockingly, the animals represented are of every age, from infant to adult. Additionally, scientist found in little fissures found "Among enormous quantities of fossilized shells, lay the two detached lower jaw and bones of the pygmy elephants and under them portions of the spinal column with ribs". Now the question Is water in the form of a flood is the agent responsible for this phenomenon?. Then HOW DID THE FLOOD GET THE POWER TO AGGLUTINATE IMMENSE ANIMALS IN FISSURES? So why the water now has no power to agglutinate enormous carcasses?.

Actually, The water being transparent, has no material properties like a hammer to nail objects. By the Same token, several elephants had been nail into the fissures in the flesh. Likewise, these kind of fossils hammer in fissures have been found in Eurasia and North and South America. The scientists stated that" These animals were not dismembered by natural predators, but by SOMETHING ACTING JUST BEFORE or DURING THEIR BURIAL on quiet literally a hemispheric acale". Obviously, a flood, water can't dismember enormous animals and hammer them in tiny fissures. or in a narrow creeks. Also, investigators found young and old animals of many incompatible species entombed together, no single individual being preserved entire and the majority in an exceedingly fragmented condition, many bones have also been found JAMMED into the furthest recesses of the caves. How water can have the power to jammed, Mammoths, Elephants and other mammals into fissures and crevices?. No the water have no power to hammer big, solid objects, Like Elephants and Mastodons. The ufo phenomenon has the technology to do those things. How the ufo phenomenon can do this unbelievable feat?.The ufo phenomenon exist at the Quantum Level and with Quantum Entanglement they can be at or three places at the same time. We all know that ufos and the aliens that fly them can become invisible. So, they neutralize the electromagnetic forces and became Invisible. We all know that aliens go through doors, walls and theirs spaceships dive into solid barriers. So, the same technology they use to jammed elephants Mastodons into fissures and crevices. They thought probably the fastest way to get rid of the evidence of their mass murder. So, the ufo phenomenon neutralize the electromagnetic Forces which is the force that creates the resistance between bodies. For instance, push agaInst the wall in you house that resistance is created by electromagnetism. Furthermore, in one cave name Choukoutien in China, scientists found large numbers of complete skeletons of incompatible animals like hyena, horse, red deer tiger, etc. In fact, it seems that the animals have been buried in the FLESH, also investigators found the fractured remains of 7 human beings SQUEEZED INTO A FISSURE . The humans represent a European, a Melanesian and skimo racial types! The scientist that found these human remains asked the question how and why such diverse racial types, of different and far apart geographical zones found themselves far removed

From their respective habitats with.

Animals of different geographical zones jammed into fissures? Yes, the Intelligence behind the Ufo phenomenon can do those things with a technology Known as QUANTUM ENTANGLAMENT. The Ufo phenomenon uses the idea in Physics of" Action at a Distance". This idea in Physics is that one body can affect another without any mechanical intervention or link between. The use of the term implies a remote and INSTANTANEOUS influence by the body without apparent mechanism for transmitting the force produced.

So, the Intelligence behind the ufo phenomenon with Quantum Entanglement can

Cover the whole earth that is the reason we see so many animals of different geographical areas in one place. By the same token, in numerous caves in Australia where there were NO ICE sheets to blame for the extinction of all the mega-fauna. In Wellington, the mega-fauna was jammed between large rocks.

The bones were in abundance, shockingly UPRIGHT! The animals showed great violence with fractures everywhere. It is clear that it was not a NATURAL DEATH. The animals were so fixed between the rocks that it was impossible to get them out.

The animals are not found in any regular position, heads, jaw, bones, teeth, ribs and femurs are jumbled and concreted together. The quantity of small animals formed deep deposits, perhaps 2 feet deep, ten wide. It must have been something prodigious "for they are compressed into the smallest possible space". Likewise, some parts of the skeleton were embedded in the cement, the ligaments still bound the bones together!

It is difficult to imagine how that could take place under any process with which we are familiar. Assuming that it was a great flood, since when the water has the property to hammer, animals between fissures? Certainly, now in the twenty first century in 2024 the water don't have these kind of property; never had it. By the same token, some animals were dismembered while still in the flesh were entombed chaotically, violently crowded unnaturally into small rock cavities and fissures. Nature as we know it don't operate like this in the extreme conditions of the Pleistocene mass extinctions. We know that even in normal emigration patterns, birds do not intermix with different species. The same with deer, bears etc. Furthermore, the Intelligence behind the ufo phenomenon use the technology of Teleportation which is the transportation of matter through space and time by converting matter into energy and then reconverting it at the terminal point. On the other hand, Quantum Entanglement which is the shifting of physical characteristics between nature's tiniest particles, no matter how apart they are in the UNIVERSE. In addition, the Intelligence behind the Ufo phenomenon is omnidirectional, which is able to receive or send radiation equally well in all directions. Also, is omnificent or unlimited in creating power.

Also, is omnifarious which has a great diversity of forms . Also, the ufo phenomenon is omnipotent which is the state of being with force of unlimited power. Also, the ufo phenomenon is omnipresent which means that it is present in all places at all times. Also, the ufo phenomenon is omniscient having infinity awareness, understanding and possessed of universal or complete knowledge. Certainly, in the Middle Ages were divine qualities now in the21 century are Technological advances. Similarly, geologist James N. Zumberg stated, "We are still unable to account for the loss of the gigantic animals of the Pleistocene. A few geologists are willing to admit quite frankly that the riddle is insoluble conundrum. A transportation of these animals has

Taken place from north to south or from east to west.

A Cave in Tasmania –

Furthermore, in a cave in Tasmania investigators found bones of four hundred opossums and two thousand different kind of mice, bats, porcupines and small birds; in total the remains of 6'888.500 animals. Actually, the stone age man it won't waste time burying mice and porcupine

When they are hungry and the Megafauna is waiting to be hunted and eaten. Also, large animals like the Megatherium, Mylodon, Toxodon,

Macrauchenia, Giant Armadillo, Jaguars. Also, Mammoths and Mastodons. Although, THEY NEVER VISITED a CAVE.

The cave of Lagoa do Sumidouro near Santa Lucia in Portugal.

Likewise, investigators found human bones intermixed with animal bones of horses, llamas capybaras etc. Also, they found the bones of over fifty human beings of both sexes and every age from infant to decrepit old man. Their skeletons lay buried in hard clay, and were discovered mix together in such confusion with other Pleistocene animals as to preclude the idea that they had been buried by an human agency. Both animal and humans had the chemical composition indicating contemporaneity

Moreover, the skull of ancient human beings were DOLICHOCEPHALIC (or long headed). A very interesting feature because according to scientists the skulls seemed to be of the old European Neanderthal race believed to be ancestral to modern Homo sapiens and to have become extinct from twenty to forty thousand years ago.

The Gypsum Cave

In this cave near Las Vegas scientists found" unusually well-preserved fresh looking remains of the camel, horse, and the mountain sheep". A great destruction of animal life took place very rapidly. The concentration in certain caves and fissures in all the continents.

The tar Pits at Mckittrick in California

By the same token, at the Tar Pits in Mckittrick, California scientists found thousands of different species and habitats of birds. The scientists asked" for what agency could have brought together in one place such dissimilar avian species?. The investigators found "Grebes, Herons Bitterns, Storks, Wood Ibises, Spoonbills, Swans, Geese, Swan Geese, Ducks, American Vultures, kites, different kind of Hawks, Falcons, Eagles, caracars, parrots, coots, ploners, stilts, sandpipers, barn owls, 7 other owls species, flycatchers, woodpeckers, swallows, jays, crows, buntings". All these birds remains were discovered by scientists in the late Pleistocene Tar Pits

At Mckittrick in California and the Asphalt Pits at Rancho La Breaa in the same state. Also, they found an intermix of birds and other animals :camel, bison, ground sloth, rodents, murrolet, black footed albatross, black vented shearwater, fummar, brandts cormorant, green winged teal, mallard duck, cinnamon teal, white fronted goose, California quail, Turquey vulture, and western Meadow lark. We see in this event how the wonderful ufo technology of Quantum Entanglement got together birds from different species, regions and latitudes brought them Together and buried them together in a common grave in the Tar Pits in California. After all, since the beginning of history there is no a single event in which a hurricane, Typhon or Tornado choose and pick people, animals or objects to take and buried.

For the simple reason that NATURE has no consciousness to pick and choose its victims. Only the Intelligence behind the ufo phenomenon can do this type of events. Similarly, at fossil graveyards at Geisel Valley, Germany

Scientists found at the lignite deposits found plants, insects and mammals. The amazing preservation of the soft parts of many of these organisms suggest a recent origin. The majority of the fossil are dismembered, it seems they have been torn apart. The silica invade the tissues must have been instantaneous because they turned into fossils. Certainly, that is the reason for preserving the membranes and original colors of the insects so fresh. The investigators stated "The plants remains are also perplexing, fungi and algae still attached to leaves impressed into the lignite. Chlorophyll is also preserved in many of the leaves which number billions. The leaves belong to plants from all parts of the world, not just from one or two climatic zones The leaves are also mostly shredded so that only their fine fibres or nervous systems remain intact. The fibres often retain their original green (chlrorophillic)color and indicated that the leaves must have been rapidly excluded with air, light and buried almost immediately they were stripped from the parent plants. Again we see how the ufo Quantum Entanglement can do all these super-feats that nature or man can't do.

Chapter 5

The Pleistocene Mass Extinction and the Strange Death of the Mammoth

Geologist have found the evidence of Glaciation cycles for the last 2'000.000 million years of at least 20 glaciation cycles with inter-glacial periods. The scientists based their studies on Green House gases preserved in ice cores taken from the Greenland Ice sheet. Also, the rise and fall in sea level is preserved in sediments deposited in coastal areas. Some of the Megafauna, the Mammoth and Mastodon they were living their lives without a problem. In fact, the Pleistocene Megafauna had no reason to disappear there is nothing written in the heavens that the natural world have to disappear. For instance, the dinosaurs lived 120 million years there was nothing written in their DNA that they have to become extinct. Again, the destruction came from the outside, an asteroid teleported by the Intelligence behind the ufo phenomenon. At least there were 4 major glaciations cycles that are well preserved in continental glacial sediments. The last major glaciation cycle ended about 11.000 to 12.000 years ago. Although, Antarctica and Greenland are both experiencing ice age conditions. The interglacial period was warmer and humans were able to develop agriculture. The climate it was much better for the surviving of the Megafauna. For instance, the Mammoth was the most successful mammal in the history of the natural world. The mammoth and the Pleistocene mammals had been through 19 previous Ice Ages, and survive. So why this end of the Pleistocene period was different? The woolly mammoth is known to science as Mammothus Primigenius and in the last

200 years very well-preserved carcasses have been discovered. According, to scientists the Mammoth appeared in Europe and Asia 40 or 50 million years ago and is member of the Elephant family. The Mammoth habitats were in North America, Canada and Alaska before becoming extinct. Mammoth carcasses had been found in Wyoming, Lake Michigan and Siberia.

In fact, in Siberia millions of mammoths remains had been found and the ivory of the giant tusks are up to 15 feet long is in perfect condition. In August 1799 an almost complete corpse of a Mammoth was found in the Delta of the Lena River. The meat was in such good condition that dogs eat the meat. Scientists began to ask why the mammoths had died out. Similarly, In August 1900 a group of hunters Found a Mammoth carcass on the banks of the Berezovka river and it was in very good condition. Moreover, carbon dating analysis of the carcass indicated the age of the Mammoth between 39.000 and 47.000 years. Shockingly, in the mouth of the Mammoth were found buttercups and grass

Interestingly, some of the plants in the stomach of the Mammoth still grow in the area. Furthermore, an analysis of the mammoths' contents revealed that contained grasses, mosses, and lichens of various kinds. Also, the green branches of tundra trees as fir and pine. Shockingly, the presence of seeds showed that the mammoth had been frozen the second half of July or August. Scientists believed that the mammoth had been browsing when stepped onto thin ice and plunged into a hole, braking its legs, and pelvis. The evidence shows that the animal suffered a very severe fall, severe enough to break his pelvis and leg. Also, large amounts of blood were found under his body. As a result of the pain his penis was found to be erect. This fact indicates that the animal was not instantly killed, but that suddenly froze to death.

Moreover, the frozen ground or permafrost might have extended down thousands of feet under surface. The permafrost or frozen ground is covered with a layer soil called muck. The muck sometimes thaw in the summer; it is composed of mud, silt and black organic matter bound together with ice. Some scientists assumed these animals fell into the ice. On the other hand, other scientist argue that there are no and never were glaciers in Siberia where the mammoths were found. Although, there is some ice in the upper slopes of few mountains. Interestingly, the animals were never found in the mountains but usually on the plains a little above sea level. Actually, all these animals never have been found in ice ; only in the muck. In addition, the animals remains there are never been found in Deltas or estuaries but in the plateaus and all over the tundra between the rivers. Interestingly, there is no evidence of GLACIERS OR AN ICE AGE IN SIBERIA, So how these animals freeze to death?. The Berezonka mammoth was discovered in a squatting position, raised up in one foreleg. The head had been mostly eaten to the bone by wolves, but much of the body was in good condition. Although, the lips and the tongue were preserved with food it had no time to swallow. Scientists found buttercups in his mouth that even today bloom in the summer . The Process of freezing. Special lists of the food industry stated that to preserve meat properly the following steps must be taken "Meat must be frozen very rapidly "if it is frozen slowly large crystals form in its cells. These crystals burst the cells and the meat begins to deteriorate at 40 Fahrenheit, it takes 20 minutes to freeze a dead turkey. Although, with the Mammoth the freezing of the flesh must be very slow enough to form crystal of ice in the cells, because the flesh deteriorate. Also, the meat of a mammoth was sampled by British scientists and it was ok. Shockingly, neither event took place with thousands of mammoths after 11.000 to 12.000 years of being frozen. Furthermore, mammoths were found in the plateaus between the river valleys. Likewise, the mammoths were fresh, whole and without damage that suggested being transported by water. The strange positions created in the death of the mammoth. For instance, A mammoth fall into the river and was carried away in an upright position and deposited hundreds of miles away. Actually, mammoths and elephants are very good swimmers Also, the great contents of fat in their stomachs, they can float for long time. So, this position of dying of the mammoth is not naturally accepted. Animals in nature don't die this way! Have you ever seeing a dog or farm animal died this way?

However, these standing mammoth with their fur coats were in perfect condition. The Berezonka mammoth it was found, "Like squatting at the back end, but was raised on one foreleg in front, with the other held forward as if about to salute. "The mammoth was perfectly frozen for their cells not too burst and not deteriorate for more than 10.000 years.! Furthermore, according to some biologists, mammoths though covered with thick underwool and a long overcoat "were not specially designed for artic conditions. For instance, their close relatives the Indian elephant, who is about the same size needs several hundred pounds of daily just to survive for more than 6 months. Certainly, there is nothing for the mammoth toeat in the tundra. Some scientists believed that "buttercups will not analysis of the contents of the stomach of the Berezonka mammoth showed a number of plants that still grow in the Artic. However, the plants today are more typical of Southern Siberia. So probably the mammoth made annual migrations north for the summer . Or that part of Siberia where the mammoths were found was in warmer latitudes. Most astonishing is that the ivory is preserved in perfect condition. However, the French dermatologist H.Neuville published in 1919 an study of the mammoth and said "Is not true that the mammoth was adapted to a very cold climate. The thick skin, hairy coat and the deposit of fat under the skin were not an adaptation to cold climate. "Also, he did a comparative microscopic study of sections of the skin of a mammoth and an Indian elephant. Indeed, discovered that were identical in thickness and in structure. The skins WERE NOT ONLY SIMILAR BUT EXACTLY THE SAME! Also, the lack of oil glands in the skin of both animals made their hair less resistant to cold and damp than the hair of average mammal. Also, he said that the common sheep is better adapted to artic climate than the mammoth. He said, that the mammoth lacks "Sebaceous glands". and that they are important for the protection against the cold. Moreover, many animals in the jungle such as lions, tigers, etc have fur which by itself does not mean an adaptation to cold. Additionally, he said, "that fur without oil is a feature of adaptation to warmth, not cold!". Moreover, there are more questions about the sudden death of

The mammoth For instance, how is possible that the body of the mammoth was deep frozen? How did it get into a solid mass of permafrost, also deep frozen, without destroying his body?. Again, I show before that the incrustations of big animals in fissures and crevices was done by the Intelligence behind the ufo phenomenon neutralizing the electromagnetic force and turning the mammoth invisible and then hammer into the permafrost.

George Cuvier : Stated "An irruption of the sea that would have carried them only from the areas where the Indian elephant now lives. . COULD NOT HAVE SPREAD them so far nor dispersed them so uniformly. Besides the inundation that buried them did not rise over the major mountain chains. For the beds that it deposited and that contains the bones are found only in low lying plains.". He proceed, "thus one cannot see how the carcasses of elephants, mammoths, mastodons etc could have been transported to the north of Siberia over the mountains of Tibet and the chains of Altai and the Urales?. "Similarly, as I show how the erratic boulders were teleported all around the world the same mammoths. Another question How did hundreds of thousands of mammoths and megafauna ended up in the Siberian Islands? Again, the mammoth and megafauna were teleported to The Siberian islands. Furthermore, the permafrost where the mammoth was found was approximately 11.000 to 12.000 years old!. According to carbon -14 test the mammoth was 39.000 to47.500 years old. The intelligence behind the ufo phenomenon with their spaceships are able to manipulate space and time don't forget the intelligence behind the ufo phenomenon exist at the quantum level.so they can manipulate space and time to create a conundrum for us to solve. Likewise, hundreds of mastodons and mammoths in the strangest positions against any common sense or natural, normal work of nature. For example, nearly all the skeletons found in the deposits in the valley of the great Osage river were in a VERTICAL POSITION! Similarly, since colonial times many mastodonts have been found STANDING ERECT! just below the surface of swamps or bogs. Likewise, the Newburgh mastodon was found, "The anterior extremities were extended under and in front of the head. The posterior extremities were extended forward under the body". Similarly, a mastodon was discovered in Monmouth County, New Jersey in 1823 it was described as follows "its vertebral column with all its joints and ribs attached to them in their natural position, lay about 8 or 10 inches below the surface the scapula rested upon the heads of the humeri and these in a vertical position, upon the forearms as in LIFE. The forearm was still buried, inclined a little backwards, and the foot which it was immediately below, it was placed in advance of the other as it would be if the animal had been WALKING!

The four feet rested on the frozen ground. Its position was vertical, the feet resting on a stratum of sand and gravel and the head to the west-south-west". Similarly, other mastodons and mammoths lying in the strangest positions.

Furthermore, a mastodon and his calf were found in an standing position in a very narrow space. The scientists that found them though that they probably were trying to reach the spring. Although, the basin was small and shallow with a bottom that was solid, solid paved with rolling stones. Once the mammoth got stuck in thick mud even a powerful animal can't turn itself upside down and become buried on its back. but not upside down!. In fact, it seems that the animals may have sensed the catastrophe because, of the strange southwest to northeast position of many of their remains. Certainly, the catastrophe was so sudden that there was not even time for these animals to be knocked over by it. So, geologically recent was their demise.

The Effects of Ufos in nature -

For instance, in 1947 in Idaho, US 10 persons witnesses 8 skimming platters in full view one of the witness A. Dishman, a housewife said, "They came in full view near St Maries, Idaho, they came in at full speed suddenly slowed and them FLUTTERED like leaves to the GROUND. The mysterious part was that we couldn't see them after they landed".

This is the best example of how the ufos dematerialize neutralize the electromagnetic forces. And they can do with erratic boulders, mammoths, etc. Next, witness reported a ufo sighting in his garden. The next day he went to check, and to his amazement the peach tree was dead! The branches and twigs were shriveled grotesquely, the leaves curled and crisp brown, and the once healthy buds of peaches looked like prunes. A little digging showed that the tree was killed to its very tap roots, overnight! The tap root is the main part of a plant growing straight downward from the stem. Similarly, trees burning black, tree branches dried and curling, trees carbonized as if petrified or turned to stone. Grass scorched, shrubbery set on fire, radioactivity levels in patches of vegetation. Sometimes the plants and vegetation grow faster than normal.

Moreover, In January, 1995 in a farm in the United States, a couple of farmers reported a ufo.

One of the farmers went early to check his livestock and 40 chickens and they found it FROZEN TO DEATH. In fact, it seems that the animals didn't panic, they were not scattered, just in their coops, DEAD. Also, three sheep were dead and had been shaved around the cheek bones, and they had a hole drilled into the cheek, all the way into the bone itself! No blood was found! The farmers stated "We cut the veins of the sheep, and there was no blood, not even in the ground. Strange tracks were found near the farm. Nobody could identify the tracks. The farmer looked at the dog, and his eyes were close.

15 minutes later, the dog was DEAD! The farmer stated "Saturday morning we went to feed. When we wake up it was already too late. All the animals were already DEAD! Likewise, the strange story of a ship named the Orange Medan, which was found wallowing in the Indian Ocean in1948 with its entire crew DEAD! And their faces FROZEN in a contortion of HORROR! Furthermore, in 1956 a night watchman at a Trenton construction firm became the first person to receive a worker's compensation claim for having been assaulted by a UFO. On the evening of October 6, 1956, the witness spotted a red light hurdling toward him. He was overcome by vile odor and collapsed with strong stomach pains. A New Jersey workman's compensation referee ruled that this extraordinary claim was legitimate and true.

The Pleistocene Mass Extinction –

In the mid- 2000s nuclear scientist Richard Firestone and colleagues presented a theory that think is correct. His idea is that a 10km wide bolide or cometary impactor struck the earth's at America sending a thermal pulse and a shock wave that resulted in widespread fires and other environmental effects which caused Mega-Faunal extinctions and caused the disappearance of the clovis culture. Also,

Cause the Younger Dryas. Also, there is no known impact of a crater. So the evidence for the event consist of various clues that they claim, point to a mass extinction created by an spatial object. The evidence included microspherules, glass melt, nano diamonds, also elevated non- earth like levels of platinum, group metals peculiar organic black mats of uncertain origin. The idea is correct but the object was a ufo like in the Tunguska explosion. the rivers attracted the Pleistocene Mammals that Went to drink water and play with their babies and meet a date. The rivers the Mammals meet and mingle. In fact, there were so many different kind of mammals. That is beyond our imagination. For instance, the short face Bear stood 13 or 14 feet Tall when all fours, it was about 25% larger than the grisly. In fact, it was the larger carnivore in North America it could run 40 miles mph.

Similarly, the ground sloth was as tall as a Giraffe and bulky as an elephant. TThe Columbian mammoth, a little smaller than the wooly Mammoth. The extinction of the Pleistocene age happened very fast. North America lost 35 species mostly large mammals, six genera or groups of species became extinct and 29 species disappeared completely from the planet. The American Lion and the dire wolf. Also, small animals such as the short -face skunk, the Aztlan rabbit and the dwarf pronghorn, which was about the size of a Golden Retriever disappear. Moreover, some scientists believed that Homo sapiens caused the extinction of the Pleistocene mammals. The Wooly Mammoth provided provisions, meat, the bones were used to build huts and the skim to cover during the very cold days. The big Pleistocene mammals were dying and disappearing in Australia, South America before the disappearing of the Mammoth, and Mastodon in North America. The Pleistocene mammals died off some suddenly and others over hundreds of years. Certainly, Homo sapiens is not responsible for the extinction of the Pleistocene mammals. For Instance, archaeological research in Clovis sites finds no signs of mammoths remains. In Pleistocene times mammals walked in fields, forests, cold, tropical, hummed, in glacial conditions etc. Pleistocene mammals were all over the world. They propagated and multiplied very healthy without any sign of diseases or premature death. The large mammals that died in North America is a long list that included Camels, Horses, Ground Sloths

The Musk-Oxen, Peccaries, Antelopes, a giant Bison with a horn spread 6 feet, a giant beaver like animal, a Stag-Moose, several kinds of cats, some of which were of Lion size, also the Imperial Elephant and the Columbian mammoth. Animals larger than the African Elephant spread all over North

America disappeared. Likewise, the mastodon that inhabited the forest and ranged from Alaska to the Atlantic coast and Mexico. Also, the woolly mammoth that roamed the areas close to the ice sheets, the Dire Wolf, the Saber Toothed tiger, the short faced bear

Additionally, the small horse disappeared and are no longer found in the old or new world. Many species of birds, these species are believed to have been the last specimens at the end of the Ice Age. As a matter of fact, the Pleistocene mammals were animals strong and healthy then suddenly died out leaving no survivors. The end came not in the struggle for existence with the survival of the fittest. Fit and unfit, old and young, with sharp teeth, with plenty of food, all perished. Furthermore, the wooly mammoth was the best example of evolution achieving perfection. The mammoth have been through 19 previous cycles of ice ages and summer and survived. His adaptation was perfect. Likewise, in Moravia eight hundred to one thousand bones were found. The shoulders blades were used in building tombs, they roamed in herds. They didn't die of starvation because they food in their stomachs when they were found. The death of the Pleistocene mammals was so sudden it seems an unexpected cataclysm fell over all the world. There was no scarcity of food. Furthermore, 200 skeletonsof mammoth were found in New York state. Moreover, fossil bones of horses which indicates that was a very common in North America. However, when Cortes arrived at the shores of America, rode their horses which they had brought from Spain. The mass extinction also wiped out the horse, in fact there were different species of horse one 3 toed Feet and one the size of a cat. The horses some went astray and went wild and filled the prairies, traveling in herds. The land, vegetation and climate were perfect for the horses. In many parts of the country fossil hunters found fossilized bones of horses in great numbers often imbedded in rock or in lava.

Furthermore, C.O Sauer had proposed the theory that the Pleistocene mammals were destroyed by man. His idea is that hunters made fires -drives in pursuit of Mammoths, Mastodons and every big game in the Pleistocene. However, is an insufficient theory to account for millions of species that disappear all around the world. Hunters burning down forests would not have been able to destroy so many millions of animals. Suddenly, disappearing and not leaving even one representative of each species alive from one coast to another and from Alaska to Tierra del Fuego. Frayne a scientist said, "In certain regions of Alaska the bones of these extinct lie so thickly scattered, that there can be no question of human work Involved". Although, because of the fast extermination of the Pleistocene mammals is impossible to say that was man to blame. Furthermore, even with maliciousness, destructiveness, evilness, of the human race armed with puny flint, tipped spears could have destroyed millions of animals and cause extinctions in all the world. L.C.Eisely wrote, "We are not dealing with a single isolated relict species but with a considerable variety of Pleistocene forms, all of which must be accorded in the light of cultural evidence an approximately similar time of extinction.". Science asks "Could it have been a disease that caused the extinction? or "the change of climate? These factors are inadequate to explain the reason why the species didn't rebound. Also, no known disease can attack so many species. Next, climate change is not enough to kill millions of animals and create a mass extinction. The extinctions in Australia were not caused by climate change neither human. However, even a sudden catastrophe all over the world can kill so many species around the world. Bot even all the volcanoes of the world erupt together would not be sufficient to destroy so many species of animals and genera. Shockingly, of some species every animal was exterminated. But no survival of the fittest they died as condemned leaving their cadavers with no signs of disease. In pits, bogs, sediments,, caverns. They died not because of lack of food, or inadequate organic evolution, or lack of adaptation. They had plentiful of food, strong bodies, perfectly adapted to the different geographical niches of Alaska –

Paleontologists ask "Under what condition did this great slaughter took place? In which millions upon millions of animals were torn limb from limb and mingled with uprooted trees remains are for the most part DISMEMBERED and disarticulate, even though some fragments still retain in their frozen state portions of ligaments, hair, skim and flesh with twisted and torn trees are piled in broken masses. Obviously, the water doesn't dismember bodies.

Assuming that was the oceans that irrupt and wash away forests with all their animal population and what caused the oceans to irrupt the continents?.

The Liakhov Islands in the Polar Circle in the Artic Ocean A hunter said, "Such was the enormous quantity of mammoth remains that it seemed that the Island was actually composed of the bones and tusks of elephants, cemented together by Icy sand". The New Siberian Islands, discovered in 1805 and 1806 as well as the Islands of Stolbonoi and Belkon present the same picture". The soil of these desolate islands is absolutely packed full of bones of elephants and rhinoceroses in astonishing numbers. These islands were full of mammoth bones, and the quantity of tusks and teeth of elephants and rhinoceroses". Did the animals come there over the ice? And for what purpose? On what food could they have live? Not on the lichens? How could large herds of mammoths, elephants in a country like Northeast Siberia, which is one of the coldest places in the world and where there was food for them? What could have caused a sudden change in the temperature of the region? Today the country does not produce for large mammals. The soil only produced moss and fungi. Furthermore, on Kotelnoi Island, there are not trees, nor shrubs, nor bushes, no fungi exist! and YET the bones of elephants, buffaloes, horses, mammoths, mastodons are found in this icy wilderness in numbers which defy all calculation. The great scientist Paul Martin believed that the Pleistocene mass extinction occurred when the arrival of Homo sapiens. Interestingly, scientists can't find evidence of mass killings of mammoths, elephants, rhinoceros, mastodons by hunters. As a matter of fact, there is no evidence that the Clovis or any other tribe left of direct contact with the mammoth and megafauna. For example, discard tools used to hunt the mammoth, mastodon and megafauna. Also, paleontologists couldn't find processing stations to cut and distributed the meat. After all, if the extinctions went so fast and so many animals were hunt where are the processing plants? For most other North American megafauna, there is no indication of interaction with humans. Likewise, hunting at such a pace should leave tools and remains of the bison or mammoth hunted. In fact, mass kill sites have never been found from the Pleistocene in North America. I am not talking about the rare Clovis point stuck in a vertebra or a bone with cut marks here and there. They are only sixteen North -American sites for which there is substantial evidence of mammoth and mastodon hunting have an age of about 12.000 to 14.500 BP which it comes to an average of one kill site every 150 years and Clovis artifacts are unknown. Furthermore, hunting large quantities of megafauna requires that the meat be deposited at a place where it can be distributed and consumed. Also, there is the problem of transportation and storage. Another reason there are no processing sites because primitive man they didn't have a social organization to better utilize surpluses of meat and storage. The disturbing reality is that for none of the well documented extinctions in the history of the earth. We don't have a solid explanation of why the extinction occurred. The Paleo-Indians of North America and the Neanderthal and all the other primitive man were not equipped to cause a mass extinction. For instance, out of 22 genera of birds that became extinct 45 percent became extinct at the end of the Pleistocene. How primitive man have the time after exterminating all the megafauna wasted his time exterminating little birds? So, the Paleo- Indians were not prepared to hunt the megafauna to extinction. They didn't have a RV to go around, they didn't have rifles, they didn't wear boots for walking in jungle, no GPS, no binoculars. So, the Paleo-Indians couldn't cause the Pleistocene Mass extinction or any other extinction. Also, according to recent research the human population may have lingered at about 1300 for more than 100.000 years and that population Bottleneck could have fueled the divergence between modern humans, Neanderthal and Denisovans. Humans might have almost gone extinct nearly 1 million ago with the world population hovering at only about 13.000 people for more than 100.000 thousand years.

Moreover, there was not a big population to eat so many millions of the megafauna and to hunt them. Anthropologist K. Kowalski points out". There is almost no evidence that primitive man was involved in the extermination of animals. "The hunting activity of primitive man, even over a very long periods do not necessarily bring the extinction of his prey. "Likewise, Sir Henry Howorth, he stated "We cannot assign the extinction of these animals to a change of climate in America, for the climate in large parts, and this where the remains most abound, has not changed at all. The same plants and the same land-shells still thrive on the same ground. The fact that the animals found were apparently in robust health when they died, with their stomachs distended with food, and the further fact that remains of many very young animals have occurred preclude the supposition that disease or want of food destroyed them. In America to the impossibility of man having caused the destruction of the animals. "Nevertheless, 40 million animals disappear .

Mass extinctions in South America –Charles Darwin was shocked when he saw the graveyards of the megafauna in South America. He wrote on January 9, 1834, in the Journal of his voyage "I t is impossible on the changed state of the American continent without the deepest astonishment. Formerly it must have swarmed with great monsters. Now we find mere pigmies". "Since they lived no very great change in the form of the land can have taken place. What, then has exterminated so many species and whole genera: The mind at first is irresistibly hurried into the belief of some great catastrophe ; but thus to destroy animals, both large and small". Interestingly, the discovery of vast quantities of animal remains in almost in every country in South America. mass Extinctions in Australia –Similarly, the mass extinctions in Australia was total with the most diverse megafauna disappearing. The Victorian fossil cave which is a warehouse for bones of more than 45.000 thousand animals

Mass extinctions in India

The Siwalik Hills are in the foothills of the Himalaya North of Delhi, they extend for several hundred miles and are two to three thousand feet high. Animal bones of species and genre living and extinct were found in the most amazing profusion. Some of the animals looked as though nature had conducted an experiment. The carapace of a tortoise 20 feet long was found. How could such animal moved in high terrain?. The animal world of today seem impoverished by comparison.   .

CHAPTER 6

THE EXTINCTION OF THE AMERICAN BISON

ꞌ The American Indian had used and eat the Bison for centuries. The Bison at the top probably numbers from 60 million to 80 million. The Indians never overhunt the Bison because they were not equip. For instance, they chase the Bison on foot because the horse disappear in the Pleistocene Mass Extinction and was re-introduced by the Spaniards. In fact, before the re-introduction of the horse, the Indians traveled on foot. Even after the re-introduction of the horse it was not common for every member of the tribe to own a horse. The Indians took turns to walk and ride. The Indians used arrow heads, spear points, axes and lances or spears. This is the reason they never can cause the extinction of the Bison. According to anthropological studies, there were probably 4 to 8 million people living in North America. The bison represented everything for the Indians in their daily life from meat, to medicines, sacred objects, robes, skins, war clubs, knives, fuel and toys. The women carved horns into cups and from the hooves made glue. Further, the bones were turned into sled runners or hoes. Also, they braided the hair into lariats or stuffed pillows. So, even with the use of very part of the Bison body. Likewise, probably by the 1680s the Indians began to ride the horse in the hunting of the bison. Next, came the introduction of firearms, gift of the white man. Before the arrival of the Buffalo hunters in the 1870s, there were sixty to eighty million Bisons in the United States. The settlers hunted the Bison for their skin sadly the meat was left to rot in a display of the greatest ignorance they didn't eat the meat. After the animal rotted, their bones were collected in large quantities and shipped east. Further, the US. Army actively endorsed the slaughter

Of the Bison; by the 1830s the Comanches, their Allies and buffalo hunters were killing about 280.000 thousand Bisons a year. The main reason for a near extinction of the Bison was the commercial market. The Bison skins were used for clothing such as robes, rugs, industrial machines belts etc. Likewise, one professional market hunter killed over 20.000 Bisons. Moreover, for a decade, there over one thousand commercial- hide markets. These commercial enterprises killed from two thousand to one hundred thousand Bisons per day depending of the season. In fact, it was said that rifles, big, 50s were fired so much that the market hunters needed at least two rifles to let the barrels cool off. However, in 1874 President Ulysses S.Grant vetoed a federal bill to protect the few thousand Bison herds, and in 1875 General Phillip Sheriday pleaded to a joint session of congress to slaughter the Bison to deprive the Indians of their source of food. Surely, by 1884 the American Bison was near extinction. As a

Result of nationwide wildlife conservation efforts initiated at the turn of the 19th century. The Bison population has surged to an estimated 250.000 they are kept in wild life reserves, Zoological parks, Yellowstone Park and private ranches. So is not new that humans caused extinctions. But never the Pleistocene Extinction If the modern Indian could barely cope with the Bison and the Grizzly bear. How did his ruder ancestor destroy the Mammoth, Mastodon, Megatherium?

CHAPTER 7

THE BLACK RHINOCEROS
EXTINCTION – DATE - 2011

1. Cuban Macawo - Extinction – Date- 1860
2. Aurosh - Extinction -Date - 1627
3. Falkland Island Wolf- Extinction – Date- 1876
4. Piranean Ibex - Extinction - Date- 2000
5. Stellers Sea Cow _ Extinction Date -1768
6. Passenger Pigeon - Extinction -Date- 1914
7. Pinta Island Tortoise - Extinction- Date -2012
8. Dodo Bird - Extinction- Date- 1662
9. Bubbal Hartbeets - Extinction -Date -1994
10. Great Auk - Extinction -Date – 1850
11. The baiji Dolphin - Extinction -Date - 2006
12. Giant Ground Sloth - Extinction -Date - 10.000 years ago
13. The Glyptodon - Extinction -Date - 10.000
14. Giant Beaver - Extinction -Date -10.000
15. The Pennsylvania Bison – Extinction -Date -1825
16. The Great Auk - Extinction -Date- 1844 -
17. The Mamo - Extinction -Date -1900
18. The Carolina Parakeet - Extinction -Date -1914
19. The Lonesome tortoise - Extinction -Date -1500's
20. The Malagasy Dwarf Hippopotamus -Date – 900s
21. The Elephant Bird -Extinction -Date – 1600s

22. The Tratratra - Extinction -date – 1500s
23. The Dodo – Extinction -Date -Before 1700
24. The Irish Elk -Extinction -Date – 10.000
25. The Woolly Mammoth Extinction - Date- 10.000
26. The European Lion- Extinction -Date – AD 100
27. The Sicilian Dwarf Elephant - extinction - date- AD 100
28. The Tasmanian Tiger -Extinction -Date - 1936
29. The Chinese River Dolphin Extinction -Date -2007
30. The Moa Extinction -Date – 1700

Scientists estimates the total number of species between 3.6 million to more than 100 million of these more than 1 million are insects. about 20.000 thousand butterfly species. About 400.000 thousand species of Beatles. The Island of Hawaii is a good example of what is happening in the world of insects, small mammals, etc. Shockingly, before humans settled the islands, they contained as many as 145 species of Birds not found nowhere else. For instance, they had native Eagles and long-legged owls also

Flightless Ibis, the Moa. Interestingly, once Hawaii had more then 10.000 different types of plants and animals. Then the Polynesian went and hunted the flightless bird to extinction. They brought the first pigs to Grassland, there were plenty of fruits banana, breadfruit and sugarcane. Today Hawaii is beautiful but no close to what it was. Of the original 145 species of birds only 35 are left and 24 are endangered. Moreover, estimates put the number of extinctions at 145.000 thousand per year.

Geological Periods and Mass Extinctions –

Archean Eon beginning of complex life - 3 Billion years ago
Protozoic Eon – Precambrian Era –(4.500 to 544 mya)
Paleozoic Era-(544 to 245 mya)
Permian –(285-245 mya) Third Mass Extinction
Carboniferous –(360-286 mya
Pennsylvanian – (325- 286 mya)
Mississippian -(360-225 mya)
Devonian –(410-360 mya) Second Mass Extinction

Silurian –(440 -410 mya)

Ordovician –(505-440 mya) First Mass Extinction

Cambrian- (544=505 mya)

Tommotian – (530=527 mya)

Mesozoic Era –(245- 65 mya)

Cretaceous – (146 - 65 – mya) Fourth Mass Extinctions

Jurassic -(208 -146 mya)

Triassic – (245 -208 mya) Fifth Mass Extinction

Tertiaria –(65-1.8 mya)

Pliocene – (5-1.8 mya)

Milocene –(23- 5mya)

Eocene- (54-38 mya)

Paleocene (65-54 mya)

Cenozoic Era Quaternary -1.8 mya to present

65mya to now

Holocene –(11.000 to now

Pleistocene –(1.8 mya to 11.000 -SIX Mass Extinction

The Pleistocene mass extinction is a legitimate mass extinction because more than 40 Million megafauna perish in North America and all over the world. This was the six mass extinction. Now we live in the seventh mass extinction that began with the Roman Empire. The Romans went to Africa and Europe and wipe out the wild life in Europe. The European Lion, the Bear, Deer etc, to sacrificated them in the Circus.

Chapter 8

Global Warming

"*These objects are conceived and directed by Intelligent Beings of a very high order ; they probably do not originate In our solar system, perhaps not even in our galaxy*".

- Dr. Hermann Oberth
German Rocket expert – 1954

"*We want the world and we wanted now*"
Jim Morrison – The Doors

Global warming is impacting the world now. The competition for natural resources specially water it will to lead to Regional Frontacions and war. Moreover, the warming in the planet is Causing the ice in Antarctica and Greenland to defrosted

Furthermore, Polar Bears cant find food and ice stand.Unfortunately, In Alaska some villages are sinking. For instance, the village of Shisumaref is sinking in to the Ocean. Similarly, in the last forty years the village of Inupiaq has lost one hundred to three hundred feet of coastline. By the same token, if the Western shelf in the Antarctic melt the oceans will rise 20 feet. As a consequence, coastal cities like New York, Florida, India, and Sri Lanka will be affected. We are at the tipping point with the climate also the melting of the Artic Ocean means a faster sea level rise and changes in winter weather, worldwide it means less rain and snow including the drought areas in the south and east. The southern Atlantic coast Georgia, South Caroline, and North Caroline will disappear beneath the ocean. The great lakes empty in the Gulf of Mexico. The Western coast of North America will be inundated, for several hundred miles inland. Similar cataclysmic disturbances will occur around the world. Norther Europe will be transformed in twinkling of an eye as land submerges and the ocean rolls inland. Also, when there is no sea ice, about 90 percent of the heat goes into the ocean which then warms everything. A vast part of the United States is warmer causing drought in the west, south east and Texas. In addition, within the last 20 years the rates of warming is about 3 times greater than the rate of warming since 1900. By the same token,

There is a water crisis in the middle east, Africa, and South America. Scientists stated that the edge of the Wilkins Ice Shelf, about the size of Connecticut is holding on by a narrow beam of thin ice and it may collapse. By the same token, if all the Western Shelf collapsed the oceans will rise to 20 feet. In addition, hundreds of thousands of animals species disappear every month, because of the destruction of nature. For Instance, frogs and bees around the world are vanishing at fast pace.

The Amazon forest, which are the lungs of the planet is disappearing at a rate of the size of England with millions of animal species. Six states Kentucky, Tennessee, South Caroline, Georgia, Alabama, Florida, had the warmest August month on record in 113 Years of record keeping. The states had to choose between agriculture and the population because of shortage of water the warmest years since we keep record are : 2023, 1998, 1997, and 1995 the second and third warmers years in record. According to recent studies global warming is happening more quickly. The planet warmed very much during the late 20 century the planet warmed 0.5F(0.25 c) a decade also, warming was greater in the Artic. Furthermore, much of Siberia has warmed by 9f(5C) 8 times faster than the rest of the planet causing melting of the permafrost, buckling of roads and toppling of buildings. The Greenland ice cap is rapidly losing thickness of artic ice. Fresh water shortages is a fact in India, China, Africa, and South America. Global warming and climate change must be reversed. According to a report heath-related deaths of people 65 and older increased by 88% between 2018 and 2022. Compared with

2000-2004 an estimated 23.200 older Americans . Mostly the burning of oil, gas and coal . The temperature is raising more quickly In the United States than the rest of the planet. Warming is intensifying wildfires in the west, droughts in the great in the Great Plains and heat waves coast to coast causing hurricanes to strenghthen more quickly in the Atlantic.

Furthermore, salt water from raising seas, is a growing threat to fresh water worldwide. Furthermore, the worlds nations need concrete steps to reverse climate change. In fact, 149 countries have updated their pledge under the 2015 Paris Climate Agreement to curb greenhouse gas emissions by 2030. The worst president in the history of our country cancelled the Paris agreement. President Biden reinstated the US in the Paris agreement. The idea is to hold the global temperature below 2 degrees Celsius or to stay at 1.5 degrees which is our actual temperature. Ecosystems are already being degraded by Habitat loss, pollution and overhunting A report said that 3 known species are becoming extinct each hour. Also, the extinction rate now is 10.000 times faster than any fossil record observed.

Conclusion

"I was in a field, a vast, grassless sad field, it did not seem to be day or night. I was walking with my brother, the brother of my Childhood, this brother of whom I must I admit I never think and whom I scarcely remember. We were talking, and we met others who were walking, we were speaking of a former neighbors, who because she lived by the street, always worked with her window open, even while we talked, we felt cold because of that open window. There were no trees in the field. We saw a man passing nearby. He was entirely naked, ashen, colored riding the color of Earth. The man was hairless, We saw his skull and the veins in his skull. He was holding a stick that was limber, like. a twig of grape vine and heavy as Iron. This Horseman passed by and said nothing. My brother said to me; "lets take the deserted road", there was a narrow, deep cut road where we saw not a bush or even a sprig of moss. All was earth colored, even the sky a few steps farther, and no one answered me when I spoke . I noticed that my bro there was no longer with me. I entered a village that I saw. I thought that it must be romainville. The first street I entered was deserted. I turned into a second street at the corner of the second street, a man was standing against the wall. I said to him "What is this place?", "Where am I?". He did not answer. I saw the open door of a house, I went in. The first room was deserted. I went into the second behind the door of this room. A man was standing against the wall. I said to this man"what is this garden?", "Where am I?" The man did not answer. I wondered through the village, and I realized it was a city. All the streets were deserted, all the doors were open. No living being was going by in the streets or moving in the rooms or walking in the gardens. But behind every turn of a wall, behind every door, behind everything, there was a man standing in silence. Only one could ever be seen at a time. These men looked at me as I passed by. I left the city and I began to walk in the fields. After a while I turned and I saw a great crowd following me. I recognized all the men I had seen in the city their heads were strange, they did not seem to be hurrying and yet they walked faster than I. They made no sound as they walked. Suddenly, came up and surrounded me their their faces were earth's colored. Then the first one I had seen and questioned as I entered he city said to me. "where are you going? don't you Know You've been dead for a long time". I opened my mouth to answer and I realized no one was there".

-Les Miserables, Victor Hugo (1802-1885)

So, it is hard to accept that extraterrestrials (the Intelligence behind the Ufo phenomenon) had been involve in Human evolution; For instance, in the 19 century it was hard to accept that something heavier than air can fly? The Airplane. Mass Extinctions had been created to advance evolution to Homo Sapiens. So, Mass Extinctions are not part of Nature and the Best example are the Dinosaurs that flourish for 120 million Years . So, the Dinousaurs are the best example that an organism left to evolve and exist will live and exist unperturbed for million of years. The idea that Homo sapiens caused the Pleistocene mass extinction is not true. Homo sapiens had no tools like all terrain trucks, GPS, rifles, no even boots to walk in the Jungle.

For instance, the Indians in the Amazons jungle never caused a mass extinction . Those populations lived in an steady hunting of the fauna before Pleistocene times replenish itself. The extinction of the Dinosaurs it caused by the ufo phenomenon. They TELEPORT the ASTEROID. Furthermore, the Intelligence behind the ufo phenomenon exist and and lived on Planet earth since during the formation of the Solar System. their nature is extraterrestrial but they lived on the planet, It seems that owned the planet. They exist everywhere in the Oceans, inside the earth, etc. They exist at the quantum level.

BIBLIOGRAPHY

Glavin, Terry. The Six Extinctions

Gould, Stephen, Jay Dinosaurs in a Haystack

Rudwick, Martin, J.S Georges Cuvier, Fossil Bones
And Geological Catastrophes

Alvarez, Walter, T Rex and the Crater of Doom

Ryan, William, Noah's Flood

Simons, Eric. Darwin Slept Here

Huxley, Julian . Evolution in Action

Montagu, Ashley, The Human revolution

Greene, John C. The Death of Adam

Howells, Willian Back to History

Sagan, Carl The Dragons of Eden

Fagan, Brian The Great Warming

Brooks, C, E, P Climate through the Ages

Coyne A. Jerry. Why Evolution is true

BIBLIOGRAPHY

About the Author

Robert Iturralde was a proud member of the United States Air Force and has been researching UFOs since 1987. The year when, by accident, he found a book about UFOs in a flea market. After reading the book, he went to the New York Times to check the reports of hundreds of witnesses; shockingly the reports were based in real eyewit- ness sightings. His previous three books are: The UFO Phenomenon and the Birth of the Jewish, Christian and Muslim Religions, A Treatise on Human Nature: Christian Saints, Historical Figures and the UFO Phenomenon, and Essay on the Theory of the Earth: Electromagnetism in UFOs and the Origin of Mass Extinctions and the Ice Ages. Mr. Iturralde is very active. He has participated in seventeen marathons, eleven triathlons and hundreds of running races. In his free time he likes to play chess and run. If you have any questions about the book you can contact him through e-mail at Gastonterryatlas@gmail.com.

Printed in the United States
by Baker & Taylor Publisher Services